£25-00

Computational Techniques
in
Heat Transfer

Computational Techniques in
Heat Transfer

Edited by:

R. W. Lewis
and
K. Morgan
Department of Civil Engineering, University College of Swansea, Wales.

J. A. Johnson
and
R. Smith
College of Forest Resources, University of Washington, Seattle, U.S.A.

PINERIDGE PRESS

Swansea, U.K.

First Published, 1985 by
Pineridge Press Limited
54, Newton Road, Mumbles, Swansea, U.K.

ISBN 0-906674-39-5

British Library Cataloguing in Publication Data

Computational techniques in heat transfer.
  1. Heat — Transmission — Mathematics
  2. Numerical calculations
  I. Lewis, R. W.
  536'.2'001511   QC320.2

ISBN 0-906674-39-5

Printed and bound in Great Britain by
Dotesios (Printers) Ltd., Bradford-on-Avon, Wiltshire

CONTRIBUTING AUTHORS

M. ALBERT — U.S. Army Cold Regions Research and Engineering Laboratory, Hanover, New Hampshire, U.S.A.

M. CROSS — School of Mathematics, Statistics and Computing, Thames Polytechnic, London, England.

R.R. EATON — Fluid Mechanics and Heat Transfer Division I, Sandia National Laboratories, Albuquerque, New Mexico 87185, U.S.A.

M. EL-NAGDY — Nuclear Engineering, Babcock Power Ltd., London, England.

K. GUSTAFSON — Department of Mathematics, University of Colorado, Boulder, Colorado 80309, U.S.A.

P.J. HEGGS — Department of Chemical Engineering, The University of Leeds, Leeds, England.

D. HITCHINGS — Imperial College, University of London, England.

M.S. HOSSAIN — Institut für Kerntechnik, Technische Universitat, Berlin, West Germany.

R.W. LEWIS — Department of Civil Engineering, University College of Swansea, Swansea, U.K.

K. MORGAN — Department of Civil Engineering, University College of Swansea, Swansea, U.K.

K. O'NEILL — U.S. Army Cold Regions Research and Engineering Laboratory, Hanover, New Hampshire, U.S.A.

A.G. PRASSAS — Management Services Department, Babcock Power Ltd., London, England.

J. RAE   Theory of Fluids Group, Harwell,
                         England.

S.T. ROBERTSON   Mechanical Engineering Department,
                         University of Lowell, Lowell,
                         Massachusetts 01854, U.S.A.

M. SAMONDS   Department of Civil Engineering,
                         University College of Swansea, Swansea,
                         U.K.

D.P. SPALDING   Computational Fluid Dynamics Unit,
                         Imperial College of Science and Tech-
                         nology, London, England.

J.F. STELZER   Kernforschungsanlage, KFA (Nuclear
                         Research Centre), 5170 Jülich, West
                         Germany.

R. SYMBERLIST   Department of Civil Engineering,
                         University College of Swansea, Swansea,
                         U.K.

V.R. VOLLER   School of Mathematics, Statistics and
                         Computing, Thames Polytechnic, London,
                         England.

L.D. WILLS   Department of Mechanical Engineering,
                         University of North Dakota, Grand Forks,
                         N.D. 58201, U.S.A.

H. WOLF   Department of Mechanical Engineering,
                         University of Arkansas, Fayetteville,
                         AR.72701, U.S.A.

CONTENTS

PREFACE

PREFACE

Recent progress in the field of numerical analysis, allied
with the rapid development of computer technology, has given
rise to major advances in the computational modelling of
various complex phenomena in almost all branches of Science
and Engineering.   In particular, the numerical simulation of
heat transfer processes has provided many notable contributions
in the evolution of both practical and efficient computer
models.   The dissemination of these achievements has been
greatly aided in recent years by the organisation of several
highly successful conferences concerned with thermal problems.
Indeed, this volume has emerged as a result of the Third
International Conference on Numerical Methods in Thermal Problems, held at the University of Washington, Seattle, in August
1983.   Following discussions during the Conference it was felt
that a more concise survey of recent developments could be
achieved by inviting a number of expanded contributions from
several of the authors present at the conference, and from
other researchers active in this area.

The first chapter by Spalding presents a comprehensive
review of non-isothermal, two-phase, flow simulation, related
to nuclear reactor systems, including current capabilities and
future trends.   Stelzer demonstrates a finite element analysis
of dry storage containers filled with spent nuclear fuel
elements.   Continuing the nuclear theme, Wills and Wolf evaluate both thermal and stress distributions within a composite
nuclear fuel rod, involving a nonlinear surface radiation
boundary condition.   The hydrological aspects of nuclear waste
storage are considered by Eaton.   Finite element models of
mass and heat transfer within a partially saturated porous
medium are employed to simulate water intrusion in rock surrounding nuclear waste repositories.

Further consideration of underground nuclear waste
chambers is given in the contribution of Hossain, where buoyancy
induced flows of leached radionuclides, in an accident scenario,
is modelled via a finite difference code.

The Rayleigh-Bénard problem of fluid motion induced by
heating within a rectangular cavity has close ties with bifurcation, or catastrophe theory, and is dealt with by Rae,
employing a finite element model.   Catastrophe theory also
plays a part in the numerical calculation, by Gustafson, of
combustion and explosion phenomena, who also introduces a
"higher-order calculus" shooting scheme.   Prassas, Hitchings
and El-Nagdy demonstrate finite element solutions of free and

forced convective flow problems. In particular, a selective reduced integration scheme is employed for incompressible systems.

A moving mesh technique, based on transfinite mappings, with deforming finite elements, is described by Albert and O'Neill, and is applied to problems involving a phase change, with associated latent heat effects. The solution of Stefan problems is also treated by Voller and Cross who employ a control volume method within an enthalpy formulation.

The coupling between heat conduction and electric current flow is modelled by Robertson using the standard Galerkin method; the problems considered having highly nonlinear properties, with phase change effects properly accounted for. Heggs deals with the extremely complex mechanisms that form part of packed bed heat transfer, which have important industrial applications.

Finally, a novel finite element scheme for dealing with mold/metal interface problems is presented by Samonds et al.

R.W. LEWIS
K. MORGAN
J.A. JOHNSON
R. SMITH

MARCH, 1985

CHAPTER 1

COMPUTER SIMULATION OF TWO-PHASE FLOWS, WITH SPECIAL REFERENCE TO NUCLEAR-REACTOR SYSTEMS

D Brian Spalding

Director, Computational Fluid Dynamics Unit
Imperial College of Science and Technology
London, SW7

ABSTRACT

A review is made of the mathematical problem of predicting the flow of interspersed phases, possessing differing (average) velocities at the same instant and location, and of successful methods of solving the problem. It is argued that, the mathematical difficulties having now been overcome, further attention should be devoted, on the computational side, to faster algorithms and more flexible computer codes; and, on the physical side, to the development of models for interfacial area, interphase transport, and interactions with solid structural elements.

1   INTRODUCTION

1.1   The purpose of the paper

The need to predict phenomena involving the flow of multi-phase mixtures, realistically and economically, is very great; and, in response, the literature of the subject has grown rapidly and in a somewhat disorderly way. As a consequence, it is hard for the by-stander to form a correct judgement as to:

● current capabilities of prediction procedures;
● the nature of the limitations to these capabilities (physical knowledge? mathematical obstacles? computer size?);
● what methods it is best to use for the time being; and
● in what direction practical research can make a useful contribution.

The purpose of the present paper is to propose answers to the questions just posed, in order that both users and researchers may see more clearly what are (in the writer's view) the most profitable directions.

Especial attention will be given, as the title of this chapter dictates, to <u>mathematical</u> aspects of the prediction procedures; and, in this regard, an effort will be made to turn the attention of the predicting fraternity away from some recent preoccupations (viz. whether the equations are hyperbolic and the problem well-posed), and towards others which may prove more fruitful.

Since the mathematical techniques are simply devices for working out quantitatively the implications of physical hypotheses, and since it is the latter that the mathematical modeller needs to be continually questioning, a part of the paper is devoted to reviewing available knowledge, and the need for more, concerning: the interphase transport processes; the interactions between multi-phase mixtures and structural elements (eg tube banks) through which they flow; and the important topic of flow-regime prediction.

1.2   The literature background

The literature of the fluid dynamics of multi-phase flow prior to 1967 is well reviewed in the book by S L Soo [1]; but, at the time of its publication, practitioners of computational fluid dynamics were still mainly concerned with solving the problems presented by <u>single</u>-phase flows.

The book by Wallis [2], which appeared in 1970, concentrated on one-dimensional flows, and especially those arising in steam/water mixtures. That the two phases present in the same region of space might be travelling at different (time-average) velocities was explicitly recognised; but the emphasis was on the provision of <u>algebraic</u> formulae, which might connect the velocity of relative "slip" with local variables such as the "void fraction" or "quality" of the mixture.

At the 1974 International Heat Transfer Conference in Japan, a "round-table discussion" was held concerning the modeling of two-phase flow, in which D Gidaspow [3] drew attention to the "complex characteristics" possessed by the set of <u>differential</u> equations which were believed to describe simple one-dimensional two-phase flows. This brought into the open a controversy that is only now subsiding. Notable contributions to the discussions were made by Francis Harlow [4] of Los Alamos Scientific Laboratories, where work on the development of numerical methods for multi-phase flows was already active.

1975 saw a major publication [5] by Harlow and Amsden. This described, and demonstrated the successful application of, a particular numerical scheme for solving the equations of motion of two inter-penetrating fluid continua. This practical success might have stilled the controversy about whether the equations did or did not present a well-posed problem; but, perhaps because other less-successful researchers were not averse to the idea of there being a "fundamental" reason for their experiences, this did not quickly occur.

The first two-phase computations by the present writer, who was luckily ignorant of what others had done in the field, whether good or bad, were made in 1976 [6]. His method proved to be similar to that of Harlow and Amsden; and the differences, which will be described below, were not such as render the computations liable to any untoward behaviour that could be attributed, plausibly or not, to ill-posedness.

Since that time, the pressure of practical need has led to the development of rather elaborate computer models of multi-dimensional two-(and more-) phase flow processes. Examples are the TRAC computer code [7], which represents the pressurised-water nuclear reactor in an accident mode, the URSULA/ATHOS computer code [8], which represents the steam-generator of the PWR system in a steady-state or operating-transient condition and various versions of PHOENICS [47] which is a general-purpose code for single- or two-phase flows.

It should be mentioned that a quite distinct approach to multi-phase fluid dynamics has been followed by other workers, notably C. Crowe [9], whose method involves the calculation of the trajectories of particles. This method, valuable though it is for flows in which the less dense phase is in large excess, cannot be employed for the majority of two-phase flows, especially those which arise in the nuclear industry. It will therefore not be considered further here.

Also ruled out of consideration, so that the scope of the paper should not be excessive, are those numerical methods [10, 11 and 12] which are applicable to flows in which the two phases are spatially separated rather than interspersed, important though these may sometimes be in nuclear systems.

In a field of science exhibiting such intense activity, the task of comprehensively reviewing the literature is a very large one; it cannot be attempted here. However, every author has a duty to perform whatever is useful when it is easy for him to do so; in this spirit, therefore, the present writer contributes the following list of his own less-ephemeral writings on the subject, as follows: [13, 14, 15, 16, 17, 18, 19, 20, 21, 22, 23, 24, 48, 49, 50, 51].

It is also proper to mention, respectfully, the recent Handbook of Multi-phase Systems, edited by G. Hetsroni [52] and two publications containing valuable survey papers namely Bergles et al [53] and Kakac and Ishii [54].

1.3   Outline of the remainder of the paper

Shortage of space (fortunately) precludes the provision of all the relevant differential equations, and a description of the various ways in which they can be averaged or otherwise manipulated; and, in any case, such topics have been well handled by others, e.g. Ishii [25] Banerjee [26] and Delhaaye [27].   Section 2, therefore, which forms a bridge to the detailed mathematical discussions of Section 3, deals with the nature of the prediction problem in general terms.   The treatment may perhaps, as a consequence, render the subject more accessible to the non-specialist; and its structure, being less cluttered with detail, may appear all the more clearly.

The mathematical analysis of Section 3 is presented in a similar "wood-before-trees" manner.   Attention is concentrated on those features that are specific to multi-phase flows; and their similarities to and differences from single-phase features will be kept in focus.   There will be no presentation or refutation of the "proofs of non-hyperbolicity of the partial differential equations"; instead, the finite-difference equations themselves will be examined in a pragmatic manner; and conclusions about the various solution procedures will be drawn.

Section 4 examines some of the terms appearing in the finite-difference equations from the point of view of their physical significance; and, on the basis of this examination, conclusions are drawn about the heavy reliance which is currently being placed upon pure guesswork concerning the physical processes involved.

This theme is taken further in Section 5 which reviews current capabilities, and Section 6 which draws attention to the need for further research and development.

2.   PREDICTION PROCEDURES FOR MULTI-PHASE FLOWS

2.1   Means of description

The central concept permitting the numerical prediction of multi-phase flows is that of space-sharing interspersed continua, according to which distinct phases (e.g. steam and water) are present within the same space (although never at precisely the same time), their shares of space being measured by their "volume fractions".

Here r and R are employed for the volume fractions of the lighter and denser of two phases, so as to permit a symmetrical representation; but the symbols α and $(1-\alpha)$ are used, for these quantities, by other authors.   As far as is appropriate, lower-case letters will here denote light-phase variables and capital letters heavy-phase ones.

According to this convention, u, v and w denote the velocity components of the light phase in the x, y and z directions; and U, V and W denote the corresponding components of the dense phase. What should be recognised at this point is that u will ordinarily differ from U, v from V, etc. <u>Two</u> fields of velocity components must be computed.

Similarily, the enthalpies of the two phases are h and H; they, too, must be computed as a rule. In general, the two phases may possess distinct pressure p and P, as when, for example the dense phase is a solid and contact between particles produces forces within them which are not shared by the fluid in the interstices.

The fundamental task of the numerical modeller is thus to devise, and to find means of solving, equations governing the ditribution through space and time of the following twelve variables:

$r, R, u, v, w, U, V, W, h, H, p, P.$

How this task can be accomplished will be described below.

If there were no limit to the length of this paper, it would be proper at this point to make explicit the assumptions which attention to the above variables (and to those only) implies. Thus, all the variables are essentially statistical; and much can be written about the alternative methods of forming the averages which they represent. Further, the restriction of attention to just two phases is neither necessary nor, in many circumstances, justified; and, indeed, the very idea of a phase can be usefully expanded, so as to permit distinctions between dense particles of the same material but different size ranges, between upward- moving and downward-moving water, etc. These topics must however be explored elsewhere.

2.2   Conservation laws

The laws of conservation of mass, momentum and energy, are obeyed by each of the phases individually. These laws generate ten partial differential equations, because there are three directions in which momentum must be balanced. Since there are twelve variables to be determined, two further equations are needed; and these are:

(i)   the "space-sharing" equation :

$$r + R = 1 \qquad\qquad\qquad\qquad\qquad ;(2.2-1)$$

and

(ii) the "particle-packing" equation, governing the pressure in one of the phases as its volume fraction approaches the physically attainable limit:

$$P = p+P^+\{R\} \qquad\qquad ;(2.2-3)$$

wherein it is the dense phase which is supposed to sustain the extra stress P+, this being a function of the volume fraction, R.

2.3   Auxiliary relations

The terms appearing in the differential equations of conservation contain:

- physical properties of the phases such as the density, viscosity, thermal conductivity, etc. ;

- inter-phase fluxes of mass, heat and momentum; and

- fictitious quantities such as the local effective (i.e."turbulent") diffusivity.

For each of these, the appropriate equations must be provided, in addition to the conservation equations, in order to permit the appropriate quantities to be deduced either from the main dependent variables r, p, h, u, etc., or from externally-supplied information.

Usually, the equations are algebraic; but, especially in the case of the quantities related to turbulent transport, additional differential equations may be appropriate, for example for the turbulence energies of the two phases, k and K, and for the length scales, l and L, of the associated eddies. "Turbulence models" of this kind [28] are well-developed only for single-phase flows, but it is to be expected that, in response to need, they will be developed for multi-phase flows also.

Some discussion of the origin of the auxiliary relations will be provided below; and it is to these that, in the writer's opinion, an increasing amount of attention should be given.

2.4   Finite-domain equations

Digital computers being capable of handling only finite sets of numbers, the partial differential equations listed above must be expressed in terms of what will here be called "finite-domain" equations. These algebraic equations, usually linear in format and connecting variables prevailing at a cluster of nodal points arranged on a grid, are most conveniently obtained by:

(i)   integrating the relevant differential equation over a domain of prescribed volume surrounding the central point of the cluster;

(ii)   evaluating the resulting volume and surface integrals with the aid of interpolation functions which are presumed to govern the distribution of the variables between the nodal points.

The presumptions about the interpolation functions are made in different ways by different workers. Unless very unwisely made, the presumptions will have no influence upon the solution when the number of grid nodes is sufficiently large; but the accuracy of the solution for a _moderate_ number of nodes _is_ influenced by the presumptions. This is a topic deserving more study than it currently receives; for numerical modelers have often been so gratified to have found a formulation which permits solutions _at all_ that they have not been eager to make further explorations.

2.5   Procedures of solution

There may easily be several thousands of nodes in the grid which is used for modeling a practical flow; and, since there are at least twelve dependent variables to be computed, some tens of thousands of simultaneous algebraic equations require solution.

Since the equations exhibit both non-linearity and coupling, iterative procedures must necessarily be employed; and these procedures are desired to "converge" with certainty, and speed, over the conditions of their employment.

Several decades of experience with single-phase flows has resulted in the development of numerous such procedures. An influential and early publication described the MAC procedure of Amsden and Harlow [29], which has subsequently been refined and improved upon [30, 31, 32]. The SIMPLE procedure of Patankar and Spalding [33] combined some of the features of MAC with some new ones, thereby giving rise to a further series of refinements and improvements [34, 35, 36].

MAC, SIMPLE, and variants, differ in the details of handling the couplings between the equations, especially those for the velocity components. Questions concerning how many variables should be up-dated simultaneously, and in what order, are answered differently by different authors, in the light of their differing needs for economy of computer time, of computer storage, and of adaptation time for new problems.

When two or more phases are present, the coupling problem is more severe than for a single phase; for the phases are bound, to some extent, to: –

(i)   interchange heat, mass and momentum by reason of their intimate contact within the same space;
(ii)   compete for occupancy of that space;
(iii) influence each other by their at-least-partial sharing of the same pressure.

These additional couplings increase the possibility of devising a plausible-seeming computational procedure which, when put to the test, does not infallibly converge, and it is such disappointments which may have contributed, wrongly in the present writer's opinion, to the concern expressed in reference [3] and subsequently elsewhere.

Section 3 of the present paper is largely devoted to the finite-domain equations which assume especial importance in multi-phase flows, and to convergent and general procedures of solving them. More than one successful procedure exists; and even a single one would suffice to demonstrate that the alarm about "well-posedness" was needless. As experience of making successful computations becomes more widely shared, the anxiety must surely abate.

2.6 Computer-code models of multi-phase processes in engineering equipment

The devising of a reliable and economical solution procedure for the multi-phase-flow equations is an essential step on the road to meeting the needs of equipment designers, operators and inspectors; but it must be followed by the construction of an equally reliable and economical computer code if these needs are indeed to be met; and it is around the computer code that the thoughts of the users often cluster.

Does TRAC properly represent the reactor re-flood process? Do ATHOS predictions agree with actual steam-generator performance? These typify the questions that are asked about actual computer codes; and it is important to recognise that they are, or may be, misleading questions. The way in which they mislead will now be explained.

A computer code is an embodiment of a solution procedure, taking in problem-specifying information, processing it in accordance with the procedure, and providing user-selected print-out of aspects of the resulting solution. The merits of computer codes should therefore be judged by reference to:-

(1)   The ease with which problem-specifying input information can be absorbed by them. This ease can be expressed in terms of:
● the variety of forms in which the information can be expressed;
● the magnitude of the amount of this information;
● the speed with which one set of input data can be replaced by another.

(2)   The economy, in terms of computer storage and time, with which a corresponding solution of the differential equation, of prescribed accuracy (not physical realism), is attained. This economy is a function of the skill exhibited by the code architect in

finding efficient ways of performing the necessary arithmetic (and avoiding that which is _not_ necessary).

(3) The convenience, comprehensibility and utility of printed-out results for which the computer code provides.

 Physical realism is of course also a merit factor of great importance; but it is so much more directly a function of the input data than of the code itself, and so greatly dependent on what questions the user asks of the computer, that statements about whether TRAC, ATHOS, RELAP, etc have been validated require much qualification before they become meaningful: the auxiliary relations which happen to have been used in the particular prediction/experiment comparisons must always be stated; for they are crucial.

 It is sometimes true that a computer code may allow only a one- or two-dimensional simulation to be made; if then it is applied to the simulation of what is in fact a three-dimensional process, the realism of the prediction will be poor. However, it is the user of the code for this purpose who should take the blame.

 Similarly, if the code does not solve for the six velocity components u, v, w, U, V, W, but embodies the "homogeneous model" $u=U$, $v=V$, $w=W$, predictions for a flow in which interphase slip is evident are bound to be poor. Once again, the code has been misused.

 These considerations show however that there is indeed a fourth merit criterion for computer codes, namely:

(4) The numbers and natures of the independent variables which can be handled. Good codes are those for which the numbers are large and the natures diverse, without the incurrence of needless expenditure or clumsiness when only few and simple variables are needed. To sum up, good computer codes are like well-equipped and well-staffed kitchens: but the quality of the meals still depends largely on the nature of the ingredients, on the recipes according to which they are combined, and on the chef.

3. A CLOSER LOOK AT THE MATHEMATICS

3.1 Introduction

 In the following sub-sections, the crucial finite-domain equations will be written down, and their implications discussed. The single-cell approach is taken, for simplicity; and alternative solution procedures are discussed with its aid.

 At first, attention is focussed on the continuity equations for the two phases; for it is these which give rise to one of the more important additional couplings. Their interactions, via the momentum equations, with the pressure (or pressures) are analysed; and it is this analysis which illuminates the key role of the "pressure correction" in the solution procedure.

Thereafter, the heat- and mass-transfer phenomena are discussed.

Much information and insight can be gained by consideration of the equations governing the pressure and volume fraction of one of the phases within a single cell of a finite-difference grid, and the flow rates through the faces of the cell.

Let such a cell have six faces, separating it from similar neighbouring cells. It will be convenient to designate the latter by the letters: N, S, E, W, H, L (standing for north, south, east, west, high, low), and the cell itself as P. There are ordinarily both mass-transport and diffusional interactions across the faces, which will be denoted by n,s,e,w,h,l; there is, in addition, a kind of "flow through time" which can be thought of as a communication across a time face with a cell denoted by P-. This "yesterday" cell represents the cell P in the state which prevailed there at the immediately precedent time step.

It will be imagined that conditions at the neighbour cells (including P-, of course) are known; to be determined are the conditions at cell P, by way of an equation which will be written as:

$$a_P \phi_P = a_N \phi_N + a_S \phi_S + a_E \phi_E + a_W \phi_W + a_H \phi_H$$

$$+ a_L \phi_L + a_{P-} \phi_{P-} + b \qquad\qquad . \ (3.1-1)$$

Of course, if the a's, and the "source term" b, are all known, the unknown ϕ_P is easily obtained. As will be seen, however, the a's are linked with the ϕ's. These linkages have been thoroughly studied for single-phase flows; but the inter-linkages are rather more complex for multi-phase flows.

Although no general guarantee can be given, a solution procedure which successfully handles the inter-linkages for a single cell will work also when applied in an iterative "Gauss-Seidel" manner, successively and repeatedly to all the cells in the grid. This is why attention will here be devoted to the manner in which equations such as (3.1-1) can be solved for single cells.

3.2   Velocity/volume-fraction linkages

(a)   Coefficients of the r and R equation

Let the volume fraction of the lighter of the two phases which are present be represented by r; and let diffusional interactions across cell faces be neglected, for simplicity. Then, when ϕ in equation (3.1-1) stands for r, the a's will represent products of cell-face areas, normal velocities, and (upwind-side) phase densities. Specifically, if u, v and w represent the velocities of this phase in the west-to-east, south-to-north and low-to-high directions, the appropriate formulae for the coefficients in the rp equation are, with < > standing for the larger of zero and the contents of the bracket:

$$a_P \equiv d_P(<-u_w>A_w+<u_e>A_e+<-v_s>A_s+<v_n>A_n$$

$$+<-w_l>A_l+<w_h>A_h+V_P/(\delta t)) \qquad \qquad ,(3.2-1)$$

$$a_N \equiv d_N<-v_n>A_n \qquad \qquad ,(3.2-2)$$

$$a_S \equiv d_S<v_s>A_s \qquad \qquad ,(3.2-3)$$

$$a_E \equiv d_E<-u_e>A_e \qquad \qquad ,(3.2-4)$$

$$a_W \equiv d_W<u_w>A_w \qquad \qquad ,(3.2-5)$$

$$a_H \equiv d_H<-w_h>A_h \qquad \qquad ,(3.2-6)$$

$$a_L \equiv d_L<w_l>A_l \qquad \qquad ,(3.2-7)$$

$$a_{P-} \equiv d_P V_{P-}/(\delta t) \qquad \qquad .(3.2-8)$$

Here the A's are the cell-face areas, V_P and V_{P-} are the cell volumes at the instant in question and at the earlier instant, and the d's are the densities of the phase in question at the locations indicated by the subscripts. δt is the magnitude of the time interval.

It should be noted that the coefficient a_P is not, as a too-superficial inspection might suggest, equal to the sum of the other coefficients; for a_P is constructed from out-flowing quantities, whereas the other a's are constructed from in-flowing ones[i].

(i) The "upwind" or "donor-cell" practice is employed here without justification or discussion. In general, the r value of fluid crossing a cell wall can reasonably be expected to lie between the limits set by the r's on the "upwind" and "downwind" sides; and to take the "upwind" value is the simplest and safest practice. A surprising implication, which however has no untoward consequences, is that the sum of the r and R of the fluid streams crossing a cell wall need not, when the two velocities have opposite signs, sum to unity.

The reason for distinguishing between V_P and V_{P-} is that it is sometimes necessary to make computations for flows in which the grid changes shape and size. The diesel-engine combustion chamber is a good example.

The quantity b, appearing on the right of equation (3.1-1), is equal to m, the mass source of the phase in question within the cell.

(b)   Compact form of the r-equation

Subsequent algebra will be simplified if the equation for r which results from the above is written as:

$$rg_o = r_i \cdot g_i + m \qquad\qquad ,(3.2-9)$$

where the subscripts o and i stand for outflow and inflow respectively, and the g's are the density-velocity-area or density-volume/time-interval quantities from which the a's are formed.   More precisely, g_o stands for the sum of all the outflows,   and $r_i \cdot g_i$ stands   for   the   sum   of   all   the volume-fraction/coefficient products of the kind: $r_n a_n$, $r_s a_s$, etc, the normally-appropriate Σ's being omitted for simplicity.   The r without subscript in equation (3.2-9) stands for r_P, the volume fraction which it is desired to determine.

Use will also be made of the corresponding equation for the volume fraction of the second phase, R;   the equation contains correspondingly-defined quantities G_o, G_i and M; and it runs:

$$RG_o = R_i \cdot G_i + M \qquad\qquad .(3.2-10)$$

If only two phases are present, the two volume fractions are linked by the consideration that, together, these two phases occupy the whole of space;   therefore:

$$r + R = 1 \qquad\qquad .(3.2-11)$$

(c)   The dependences of the g's and G's on the pressure within the cell

Let it first be supposed that a single pressure suffices to characterize conditions in the cell.

A rise in pressure within the cell tends to increase the outflow velocities, and to diminish the inflow ones.   The g's and G's are therefore sensitive to cell pressure;   and, if only small variations   are   considered,   the   relationships   can   usually   be represented by way of the linear equations:

$$g = g^* + g'p' \qquad\qquad ,(3.2-12)$$

$$G = G^* + G'p' \qquad\qquad ,(3.2-13)$$

wherein the starred quantities are the values of g and G which prevail when the cell pressure has some guessed value, and p' represents the correction to this pressure which must be applied in order that all the conservation equations (ie those for the masses of the two phases, and for their momentum fluxes at the six cell faces) are to be satisfied.

By way of clarification, it may be mentioned that:-

- g'$_i$ and G'$_i$ are always negative, because a pressure rise dimishes the rate of inflow;
- g'$_o$ and G'$_o$ are always positive, because a pressure rise increases outflow;
- the g-p' relation is, strictly speaking, <u>non</u>-linear; for g$_o$ cannot become negative without ceasing to be an outflow, nor g$_i$ without ceasing to be an inflow;
- the g'$_i$ corresponding to "flow across the time face" is zero, because a change in late-time pressure cannot bring about a difference in early-time density.

(d)   The pressure-correction equation

By combining the equations (3.2-9), (3.2-10) and (3.2-11) with the expressions for the g's, the following equations can be derived:

$$\frac{r_i \cdot (g_i{}^* + g_i 'P') + m}{g_o{}^* + g_o 'P'} + \frac{R_i \cdot (G_i{}^* + G^i 'P') + M}{G_o{}^* + G_o 'P'} = 1$$

.(3.2-14)

This equation, from which the unknown volume fractions r and R have been eliminated, and in which the r$_i$ and R$_i$, valid for neighbour points, are known, has the following important properties:-

- It is a (superficially) quadratic equation for the sought-for pressure correction, p'; so it may be solved by way of the usual quadratic equation formula.

- Because g'$_i$ and G'$_i$ are negative, while g'$_o$ and G'$_o$ are positive, and because neither the numerators nor denominators in equation (3.2-14) can ever be negative, the left-hand side varies smoothly and monotonically from positive infinity at a suitably small p' to zero at a suitably large p'; it therefore always equals the right-hand side, i.e. unity, at a particular finite value of p', i.e. the solution.

- The equation consequently always has a unique physically realistic solution; the second solution to be expected of a quadratic is precluded by the g-not-less-than-zero discontinuity mentioned above.

3.3 Procedures for solving the single-cell one-pressure problem

(a) Order of events

The following sequence of operations is now to be envisaged, with the aim of determining values, satisfying all the relevant equations, of the following:

● $p, r, R, d (\equiv$ light-phase density) and $D (\equiv$ heavy-phase density) at the point P;

● u, v, w, U, V, W at the cell faces (i.e. twelve velocities in all).

The sequence is:-

(i)   Guess the pressure p. Call this p^*.

(ii)   Compute values of the twelve velocities from the appropriate momentum equations, corresponding to p^*.

(iii)   Compute the values of d and D which correspond to p^*, with the aid of the pressure-density relations of the phases.

(iv)   Compute the corresponding values of the g and G quantities, call these g_o^*, g_i^*, G_o^*, G_i^*.

(v)   Substitute these values into equation (3.2-14) and solve for p'.

(vi)   Solve for g and G from equations (3.2-12) and (3.2-13)

(vii)   Substitute into equations (3.2-9) and (3.2-10) to yield r and R.

(viii) Add p' to p^* to yield p.

(ix)   Return to step (ii) and repeat the process, until the values of the pressure corrections are acceptably small.

Iteration is necessary for a variety of reasons, namely:-

● the coefficients of the velocity equations themselves depend upon the velocities, volume fractions and r's that are being computed;

● the velocity pairs u_w and u_e, v_s and v_n, w_l and w_h, V_w and U_e, etc. are linearly linked together in a way that has probably not been accounted for in their respective equations;

● the velocity pairs u_w and V_w, u_e and U_e, v_s and V_s, etc., are linked together (by inter-phase friction) in a way that is probably non-linear, and may not have been accounted for even in a manner appropriate to linear linkage (see section 3.5 (a) below).

● the pressure-density relations may be non-linear ones.

This general necessity to iterate needs to be taken into consideration when the question of how to solve equation (3.2-14), in step (v) is decided.

(b) **Use of the quadratic formula for r**

Equation (3.2-14) can be oxpressed in the form:

$$\alpha p'^2 + \beta p' + \lambda = 0 \qquad\qquad ,(3.3-1)$$

by the substitutions:

$$\alpha \equiv r_i \cdot g_i 'G_o' + R_i 'g_o' - g_o 'G_o' \qquad\qquad ,(3.3-2)$$

$$\beta \equiv (r_i \cdot g_i^* + m) G_o' + G_o^* (r_i \cdot g_i ' - g_o')$$

$$+ (R_i \cdot G_i^* + M) g_o' + g_o^* (R_i \cdot G_i ' - G_o') \qquad\qquad .(3.3-3)$$

$$(r_i \cdot g_i^* + m) G_o^* + (R_i \cdot g_i^* + M) g_o^*$$

$$- g_o^* G_o^* \qquad\qquad .(3.3-4)$$

Then the solution can be found by way of the formula:

$$p' = \frac{-2\lambda}{B + \sqrt{B^2 - 4\alpha\lambda}} \qquad\qquad ,(3.3-5)$$

whereby the positive root has been taken for physical reasons.

Equation (3.3-5) may represent the desired solution for the pressure correction p', from which values for the g's and G's, and for r and R, can be derived. However, before accepting this value of p'can be accepted it is necessary to check all values of $g^* + g'p"$ and $G^* + G'p'$; for, if any proves to be negative, it must be removed from the outflow group and put into the inflow group, or vice versa; then the equation should be solved again.

(c) **The linearisation procedure**

Partly because of the awkwardness of the procedure just described, partly because cell-by-cell adjustments in multi-dimensional computations are rare, and partly because iteration is needed in any case, it is usually regarded as preferable to linearise the procedure in some way, for example by neglecting the p'^2 term in equation (3.3-1), with the result:

$$p' = \frac{-\lambda}{\beta}, \text{ i.e.}$$

$$p' = \frac{g_0^* G_0^* - (R_i \cdot g_i^* + m)G_0^* - (R_i \cdot g_i^* + M)g_0^*}{\{(r_i \cdot g_i^* + m)G_0' + (R_i \cdot G_i^* + M)g_0' + G_0^*(r_i \cdot g_i' - g_0'}$$

$$+ g_0^*(R_i \cdot G_i' - G_0')\} \qquad .(3.3-6)$$

The significance of this procedure can be best seen by defining r^* and R^* as those values of the volume fractions which would correspond to the "starred" mass flows, i.e. by:

$$r^* \equiv (r_i \cdot g_i^* + m)/g_0^* \qquad .(3.3-7)$$

$$R^* \equiv (R_i \cdot G_i^* + M)/G_0^* \qquad .(3.3-8)$$

With these substitutions, and some easy manipulations, equation (3.3-6) becomes:

$$p' = \frac{(r^* + R^* - 1)}{\{(1 - R^*)g_0' - r_i \cdot g_i'\}/g_0^* + \{(1 - r^*)G_0' - R_i \cdot G_i'\}/G_0^*} \qquad ;(3.3-9)$$

and, since $1 - R^*$ must be very nearly r^*, and $1 - r^*$ must be very nearly R^*, this can be even more neatly expressed as:

$$p' \approx \frac{(r^* + R^* - 1)}{(r^* g_0' - r_i \cdot g_i')/g_0^* + (R^* G_0' - R_i \cdot G_i')/G_0} \qquad .(3.3-10)$$

Inspection of this equation permits the following deductions to be made: –

- The numerator must be zero, when the true solution is attained; thus, as is to be expected, the magnitude of the pressure correction p' is proportional to the magnitude of the error which is to be suppressed.

- Because g_i' and G_i' are never positive, and because g_o' and G_o, are never negative, the denominator is always positive. This means that, when the sum of r^* and R^* exceeds unity (i.e. more material is being pushed into the cell than can be accommodated there), the pressure correction must be positive. This is obviously appropriate; for an enlarged pressure in the cell will increase the rates of outflow and decrease the rates of inflow.

- Should the equation be applied in a single-phase region, in which r_i is unity and R_i and M zero, the equation, in the form of (3.3-6), reduces to:

$$p' \approx \frac{g_i^* + m - g_o^*}{g_o' - g_i'} \qquad\qquad .(3.3-11)$$

This is precisely the same as the pressure-correction equation that is employed in the SIMPLE procedure of Patankar and Spalding (33), developed, as described above, for single-phase flows. This result is expected.

(d)   Successive-substitution procedures.

Some computational economy, albeit accompanied by loss of symmetry as between phases, arises when the quantity R^* is deduced from equation (3.3-8), whereupon $1 - R^*$ is substituted for r in the equation of total-mass conservation, formed by adding equations (3.2-9) and (3.2-10). There results:

$$(1 - R^*)g_o + R^*G_o = r_i.g_i + m + R_i.G_i + M \qquad\qquad .(3.3-12)$$

After some algebra, and again the setting of r^* to equal $1 - R^*$ in the denominator of the expression, there results:

$$p' \approx \frac{(r^* + R^* - 1)}{(r^*g_o' - r_i.g_i')/g_o^* + (R^*G_o' - R_i.G_i')/g_o^*} \qquad\qquad .(3.3-13)$$

This equation would be used in place of (3.3-6), in the iteration loop, so it might be regarded as solving the quadratic by successive substitution.

Inspection of this equation, and comparison with equation (3.3-10), permits the following observations:-

- Although similar in general form, the equations differ in that the second term in the denominator is now divided by g_0^* rather than G_0^*. This difference can be very great, for, if the velocities of the two phases are very similar G_0^*/g_0^* will be approximately equal to the ratio of the densities of the two phases.

- Had the process been "unsymmetrical" in a different way, beginning with the substitution of $1-r^*$ for R in the total-mass conservation equation, the result would have been:

$$p' \approx \frac{(r^*+ R^*- 1)}{(r^*g_0'-r_i.g_i')/G_0^* + (R^*G_0'-R_i.G_i')/G_0^*}$$

$$,(3.3-14)$$

which differs from equation (3.3-10) in the divisor of the <u>first</u> term in the denominator. This value of p' differs yet again, and perhaps very considerably, from both the earlier ones.

- Obviously, not all these p''s can be the best ones for the promotion of rapid convergence to the solution. Since the linearisation procedure of equation (3.3-6) is exact for small p', and there is no reason to suppose that equation (3.3-13) and (3.3-14) are more correct for large p', it is equation (3.3-6) that is to be recommended.

(e) <u>The use of the volumetric continuity equation</u>.

A successive-substitution procedure that has been employed in some computer codes by the present writer combines the unsymmetrical choice of one or other of the volume fractions with the equation which is formed by <u>dividing equations (3.2-9) and (3.2-10) by the relevant fluid density prior to addition.</u> This equation is:

$$d^{-1}(rg_0-r_i.g_i-m) + D^{-1}(RG_0-R_i.G_i-M) = 0$$

$$.(3.3-15)$$

If R^* is substituted in (3.3-15) for R, $1-R^*$ for r, $g^*+g'p'$ for g and $G^*+G'p'$ for G, the result is:

$$d^{-1}\{(1-R^*)(g_0^*+g_0'p')-r_i.(g_i^*+g_i'p')-m\}$$

$$+D^{-1}\{R^*(G_0^*+G_0'p')-R_i.(G_i^*+G_i''p')-M\} = 0$$

$$.(3.3-16)$$

Now R^* satisfies equation (3.3-7); and r^* can be introduced from equation (3.3-8). The result is:

$$p' = \frac{(R^* + R^* - 1)}{\{(-R^*)g_0' - r_i \cdot g_i'\}/g_0^* + (d/D)(R^* G_0' - R_i \cdot G_i')/g_0^*}$$

.(3.3-17)

Finally, by setting $(1.R^*)$ approximately equal to r^*, one may deduce:

$$p' \approx \frac{(r^* + R^* - 1)}{(r^* g_0' - r_i \cdot g_i')/g_0^* + (d/D)(G_0^*/g_0^*)(R^* G_0' - R_i G_i')/G_0^*}$$

.(3.3-18)

An alternative equation can be derived by making the other "unsymmetrical" choice of substituting r^* for $(1-R^*)$ and r^* for r in (3.3-15); it is:

$$p' \approx \frac{\langle r^* + R^* - 1)}{(D/d)(g_0^*/G_0^*)(r^* g_0' - r_i g_i')/g_0^* + (R^* G_0' - R_i G_i')/G_0^*}$$

.(3.3-19)

Comparison of equations (3.3-18) and (3.3-19) with that derived by linearising the quadratic, namely (3.3-10), shows that they differ by way of the (variously-placed) factor $(d/D)(G_0^*/g_0^*)$. This factor, whenever the velocities of the two phases are equal, is identically equal to unity; and, since the velocities often are not far from equality, the factor will often be close to unity. It is this fact, presumably, which explains the success which has attended the use of the volumetric-continuity equation.

(f)   The recommended pressure-correction equation

Within reasonable limits, it matters little what precise form of pressure-correction equation is employed. What is desired is to shift the pressure in the right direction, and as closely as possible by the right amount; but the non-linearities and inter-linkages preclude, in any case, the attainment of the exact solution in a single move. This is why substitutions of, for example, r^* for $1-R^*$ have been allowable; and it explains why the uses of (3.3-18) and (3.3-19) have been successful, even when $(d/D)(G_0^*/g_0^*)$ has taken values rather different from unity.

This insensitivity to the precise form of the denominator of the p'-expression permits equation (3.3-10) to be further simplified, with little loss of accuracy and much gain of simplicity, to:

$$p' = \frac{r^* + R^* - 1}{r^* g'/g_0^* + R^* G'/G_0}$$

.(3.3-20)

Here the symbol g' is being used for the sums of all the g_o'''s; and G' has a similar significance. For the sake of certainty, an expression for g' will be written out in full for the cartesian cell described at the beginning of sub-section 3.2; it is:

$$g' = d_p \left\{ -\frac{\partial u_w}{\partial P_p} A_w + \frac{\partial u_e}{\partial P_p} A_e - \frac{\partial v_s}{\partial P_p} A_s + \frac{\partial v_n}{\partial P_p} A_n \right.$$
$$\left. -\frac{\partial w_l}{\partial P_p} A_l + \frac{\partial w_h}{\partial P_p} A_h \right\} + \frac{\partial d_p}{\partial P_p} \cdot \frac{V_p}{\delta t}$$

$$;(3.3-21)$$

and there is a similar expression for G'.

(g) The recommended p' expression when simultaneous adjustments are made to neighbouring pressures.

It is now convenient to recall that, in reality, single-cell problems such as have been considered here arise, practically, in multi-cell contexts: the adjustments which are being made to the pressure in a particular cell are accompanied, as a rule, by adjustments to the pressures at neighbouring cells also.

This fact can be reflected by noting that equations (3.3-12) and (3.2-13) must be replaced by:

$$g = g^* + g'(p'-p'')$$,(3.3-22)

$$G = G^* + G'(p'-p'')$$,(3.3-23)

where p" stands for the correction to the neighbour-cell pressure.

Carrying such amendments through to the limit leads to the derivation of an algebraic equation which connects p'_p with the pressure corrections at neighbouring points, namely: P'_N, P'_S, P'_E, P'_W, P'_H, P'_L. It can be expressed in a form similar to (3.1-1), namely as:

$$a_p p' = a_N p'_N + a_S p'_S + a_E p'_E + a_W p'_W$$
$$+ a_H p'_H + a_L p'_L + b$$

$$;(3.3-24)$$

and the corresponding definitions of the a's and b are:-

$$a_N \equiv A_n \left(\frac{\partial v_n}{\partial P_p} \cdot f + \frac{\partial v_n}{\partial P_p} \cdot F \right)$$,(3.3-25)

$$a_S \equiv -A_s \left(\frac{\partial v_s}{\partial P_p} \cdot f + \frac{\partial v_s}{\partial P_p} \cdot F \right)$$,(3.3-26)

$$a_E \equiv A_e \left(\frac{\partial u_e}{\partial P_p} \cdot f + \frac{\partial U_E}{\partial P_p} \cdot F \right)$$,(3.3-27)

$$a_W \equiv -A_w\left(\frac{\partial u_w}{\partial p_P} \cdot f + \frac{\partial U_w}{\partial p_P} \cdot F\right) \qquad ,(3.3-28)$$

$$a_H \equiv A_h\left(\frac{\partial w_h}{\partial p_P} \cdot f + \frac{\partial W_h}{\partial p_P} \cdot F\right) \qquad ,(3.3-29)$$

$$a_L \equiv -A_l\left(\frac{\partial w_l}{\partial p_P} \cdot f + \frac{\partial W_l}{\partial p_P} \cdot F\right) \qquad ,(3.3-30)$$

$$a_P \equiv a_N + a_S + a_E + a_W + a_H + a_L + \frac{V_P}{\delta t}$$

$$\left(\frac{1}{d_P} \cdot \frac{\partial d_P}{\partial p_P} \cdot f + \frac{1}{D_P} \cdot \frac{\partial D_P}{\partial p_P} \cdot F\right) \qquad ,(3.3-31)$$

$$b \equiv r^* + R^* - 1 \qquad ;(3.3-32)$$

and the quantities f and F, appearing above, are defined by:

$$f \equiv r_P / \{A_n\langle v_n\rangle + A_s\langle -v_s\rangle + A_e\langle u_e\rangle + A_w\langle -u_w\rangle$$

$$+ A_h\langle w_h\rangle + A_l\langle -w_e\rangle + V_P/\delta t\} \qquad ,(3.3-33)$$

$$F \equiv R_P / \{A_n\langle V_n\rangle + A_s\langle -V_s\rangle + A_e\langle U_e\rangle + A_w\langle -U_w\rangle$$

$$+ A_h\langle W_h\rangle + A_e\langle -W_e\rangle + V_P/\delta t\} \qquad .(3.3-34)$$

The expressions for r^* and R^* will also be written out in detail; for there is less latitude, indeed none, in the way in which b is calculated. The equations to be used, again for the cartesian cell, are:

$$r^* \equiv \left(\frac{f}{d_P r_P}\right) \{d_N r_N A_n\langle -v_n\rangle + d_S r_S A_s\langle v_s\rangle$$

$$+ d_E r_E A_e\langle -u_e\rangle + d_W r_W A_w\langle u_w\rangle$$

$$+ d_H r_H A_h\langle -w_h\rangle + d_L r_L A_l\langle w_l\rangle$$

$$+ d_{P_-} r_{P_-} V_{P_-}/\delta t + \dot{m}\} \qquad ,(3.3-35)$$

$$R^* \equiv \left(\frac{F}{D_P R_P}\right) \{D_N R_N A_n\langle -V_n\rangle + D_S R_S A_s\langle V_s\rangle$$

$$+ D_E R_E A_e\langle -U_e\rangle + D_W R_W A_w\langle U_w\rangle$$

$$+ D_H R_H A_h\langle -W_h\rangle + D_L R_L A_l\langle W_l\rangle$$

$$+ D_{P_-} R_{P_-} V_{P_-}/\delta t + \dot{M}\} \qquad .(3.3-36)$$

Equation (3.3-24), with equations (3.3-25) through (3.3-36) which define the terms appearing in it, occupies a central place in procedures for the numerical computation of multi-phase flows. It represents the "Poisson equation for pressure correction", more elementary forms of which have been central to such single-phase prediction procedures as SIMPLE (37). It is recommended for use for all single-pressure problems.

However, two-pressure problems also arise. These will be discussed in the following section.

3.4   Solution procedure for a two-pressure problem

(a)   Origin

It has already been mentioned, in connexion with equation (2.2-3), that an additional pressure, P^+, may act upon the denser phase. There are two common circumstances in which this occurs, namely:

(i)   the dense phase is a powder, or granular solid; and, when this is closely packed (R becomes large), contact between grains provokes a pressure resisting further compaction;

(ii)   the dense phase is a liquid, and, because the dimensionality of the process has been artificially reduced with neglect of the vertical momentum terms, the effect of gravity is being accounted for by inclusion of appropriate sources in the horizontal momentum equation.

In both these circumstances, P^+ tends to increase with R. In case (i), P^+ is zero until R reaches a critical value, and then rises rapidly; in case (ii), P^+ increases more gradually (often linearly) with R. The differential coefficient $\partial R/\partial P^+$ has significance similar to that of compressibility.

Let $\partial R/\partial P^+$ be given the symbol $R^{+\prime}$. Its magnitude diminishes with the rigidity of the phase in circumstance (i), above, and with the gravitational acceleration in circumstance (ii). In section 3.3, $R^{+\prime}$ was implicitly taken to be infinite; the task is now to determine how the solution procedure should be modified to account for finite or zero $R^{+\prime}$.

(b)   The modified solution procedure

The order of events described in sub-section 3.3(a) now requires to be modified, as follows: -

(i) P^+ must also be guessed. Let its value be P^{+*}; and let the dense-phase volume fraction to which it corresponds be denoted by R^{+*}.

(ii), (iii), (iv)   These steps are carried out with p^*+P^{+*} in the place of p^* in the dense-phase momentum equations, and in the calculation of D.

(v)   a different equation from (3.3-20) must be solved for p'; and an increment, $P^{+'}$, must also be computed for P^+.   Details will be given in sub-section (c).

(iv)   through (ix)   These are as before, or are modified in obvious ways.

(c)   The pressure-correction equations

The single-cell approach will be taken, for simplicity.

Equation (3.3-20), which expresses the connexion between the flow-rate changes and the continuity error which is to be eliminated, must be modified, to accord with the considerations of sub-section (b), to:

$$p'r^*g'/g_0{}^* + (p'+P^{+'})\ R^*G'/G_0{}^* = r^* + R^* - 1$$

$$.(3.4-1)$$

The second-pressure correction, $P^{+'}$, can be related to p' in another way.   This is the requirement that the modified value of R resulting from the flux changes, namely:

$$R = R^* - (p'+P^{+'})\ R^*G'/G_0{}^* \qquad ,(3.4-2)$$

must also accord with the P^+/R relation representing the "extra rigidity" of the dense phase, viz:

$$R = R^{+*} + P^{+'}\ R^{+'} \qquad .(3.4-3)$$

Algebraic manipulation now permits derivation of equations for p' and $P^{+'}$ independently of each other.   They are:-

$$p' = \frac{R^{+'}(r^*+R^*-1)+(R^*G'/G_0{}^*)(r^*+R^{+*}-1)}{R^{+'}(r^*g'/g_0{}^*+R^*G'/G_0{}^*)+(R^*G'/G_0{}^*)(r^*g'/g_0{}^*)}$$

$$.(3.4-4)$$

and:

$$P^{+'} = \frac{(r^*g'/g_0{}^*+R^*G'G_0{}^*)(R^*-R^{+*}) - (R^*G'/G_0{}^*)(r^*+R^*-1)}{R^{+'}(r^*g'/g_0{}^*+R^*G'/G_0{}^*)+(R^*G'/G_0{}^*)(r^*g'/g_0{}^*)}$$

$$.(3.4-5)$$

Also useful is the relation for the sum of these quantities, namely:

$$p'+P^{+'} = \frac{R^{+'}(r^*+R^*-1) + (r^*g'/g_0{}^*)(R^*-R^{+*})}{R^{+'}(r^*g'/g_0{}^*+R^*G'/G_0{}^*) + (R^*G'/G_0{}^*)(r^*g'/g_0{}^*)}$$

.(3.4-6)

As is to be expected, the difference between R^{+*} and R^* now plays a part in determining the pressure corrections, not just the difference between r^*+R^* and unity.

(d) Discussion

The mathematical nature of the problem is not significantly changed by the two-pressure complication, as may be recognised from the following considerations of special cases:

(i)   When $R^{+'}$ is very large, which implies that large changes in R can take place without causing P^+ to change significantly, equation (3.4-4) reduces to (3.2-20), as is to be expected. The second pressure can be neglected in these circumstances.

(ii)   When $R^{+'}$ is very small, on the other hand, equation (3.4-4) reduces to:

$$p' = \frac{r^* + R^{+*} -1}{r^*g'/g_0{}^*}$$

.(3.4-7)

This implies that the light-phase pressure correction is to be computed independently of what is happening to the dense phase; for R^* has disappeared from the numerator and $R^*G'/G_0{}^*$ from the denominator.

In these circumstances (small $R^{+'}$), the expression for $p'+P^{+'}$ in equation (3.4-6) becomes especially simple. It reduces to:

$$P' \equiv p' + P^{+'} = \frac{(R^*-R^{+*})}{R^*G'/G_0{}^*}$$

.(3.4-8)

This implies that, in its turn, the dense-phase pressure correction, P', is to be computed by reference to the dense-phase terms only; this confirms that the two phases are, in effect, influenced by two almost independent pressure fields.

(iii)   The dimensionless quantity which determines the nearness of a given situation to one extreme or the other is $R^{+'}/(R^*G'/G_0{}^*)$. The smaller this is, ie the more "rigid" is the second phase, the more different the two pressure fields become.

(iv)   However, the solubility of the equations is in no way diminished by their additional complication.   No shadow of "ill-posedness" is detectable;   and this remains true when the equations are extended to represent also the situation in which the pressures in neighbouring cells are also changing. This extension will not be presented here, for lack of space.

(v) The practitioner of numerical fluid mechanics learns to distrust confident statements, of the kind just made, when they are not supported by practical experience. In the present case, that support can be provided: many computations by the author and his colleagues have confirmed the robustness of the solution scheme; and nothing to suggest ill posedness has been encountered.

3.5 Interphase transfer processes

(a) Momentum transfer

Because of the friction which takes place between the two interspersed phases, their momentum equations are inter-linked in a manner of which account needs to be taken, if the solution procedure is to be economical. The problem, and how it is solved, will be briefly indicated by reference to just two velocity components: u_e and U_e, say.

The finite-domain momentum equations governing them can be expressed, if only immediately interesting components are brought into prominence, as:

$$l\{u_e,\ p_P - p_E,\ U_e\} = 0 \qquad\qquad ,(3.5-1)$$

and

$$L\{U_e,\ p_P - p_E,\ u_e\} = 0 \qquad\qquad ,(3.5-2)$$

where $l\{\ \}$ and $L\{\ \}$ stand for linear functions. U_e appears in equation (3.5-1), and u_e in equation (3.5-2), because of interphase friction and interphase mass transfer.

Since the equations are linear, elimination of u_e and U_e from each other's equations is easy. There result further linear equations which can be written as:

$$l'\{u_e,\ p_P - p_E,\ p^+_P - p^+_E\} = 0 \qquad\qquad ,(3.5-3)$$

and

$$L'\{U_e,\ p_P - p_E,\ p^+_P - p^+_E\} = 0 \qquad\qquad ,(3.5-4)$$

where l' and L' are new linear functions.

What is important is that u_e and U_E, and the differential coefficients $\partial u_e / \partial p_P$ and $\partial U_e / \partial p_P$, which appear in the coefficients a_E and a_P of sub-section 3.3(f), should be obtained by differentiation of the second pair of equations rather than the first;

26

for, otherwise, extremely slow convergence may result when the interphase friction coefficient, or the mass-transfer rate, is high. This "hobbling" phenomenon has been described by the present writer in more detail elsewhere (6,8).

(b) Heat transfer

The enthalpy equations for the two phases, in a finite-domain form which exposes all that is needed for the present discussion, can be written:

$$1\{h_P, q, h_i\} = 0 \qquad\qquad ,(3.5\text{-}5)$$

and

$$L\{H_P, Q, H_i\} = 0 \qquad\qquad .(3.5\text{-}6)$$

Here q and Q stand for heat transfers into the phase from elsewhere (eg an interspersed solid, such as the fuel rods of a nuclear reactor, or the tubes of a heat exchanger); and h_i and H_i stand for the enthalpies of the phases at the interface between the two phases.

It is now useful to distinguish two extreme cases. In the first, which is the one which is commonly encountered in the nuclear industry, the two phases are the vapour and liquid of a single chemical substance; they are in intimate contact; and the pressure is high enough for thermodynamic equilibrium to prevail at the interface between them. The quantities h_i and H_i can then be replaced by h_{sat} and H_{sat}, the "saturation" values appropriate to the prevailing pressure.

In the other extreme, the two phases may have differing chemical compositions; and the only thing in common between the two interface conditions is their shared temperature. It then becomes useful to replace h_i and H_i in the linear expressions by the interface temperature.

Nothing more will be said about the second extreme case than that it often necessitates, for efficient solution, an application of the partial-elimination technique which was described in sub-section (a) in respect of momentum transfer; and this remark is itself made only to introduce the statement that no such partial elimination is advantageous for the first extreme, which alone will be discussed further.

In the nuclear-reactor (steam-water) situation, a major role of equations (3.5-5) and (3.5-6) is to permit the computation of the interphase mass-transfer rate. This is deducible from some such equation as:

$$m = -M = \frac{\{c(h_P-h_{sat}) + C(H_P-H_{sat})\}}{\lambda} \qquad ,(3.5\text{-}7)$$

wherein c and C represent phase-to-interface conductances per unit volume (which also make their appearances in equations (3.5-5) and (3.5-6) respectively), λ is the latent heat of phase change, and it has been presumed that the source of mass in the vapour phase is equal to the sink of mass in the liquid phase.

It is true that the interphase mass-transfer rate appears also directly in equations (3.5-5) and (3.5-6); but its role there is subordinate, and successive guesses and corrections, which fit easily into the general framework of solution, handle effectively the associated non-linearity.

In some models of two-phase flow, it is supposed that c and C are so large that h_P and H_P can never be different from h_{sat} and H_{sat} respectively, whenever both phases are present simultaneously. This is the case with the early version [8] of the URSULA code, for example; but later versions have the unequal-temperature capability. No difficulty is experienced in solving the equations, whatever model is employed.

(c) Mass transfer

One commonly-valid relation for the interphase mass-transfer rate has just been given, as equation (3.5-7); but there are two others, which will be briefly mentioned.

(i) When the vapour phase contains an incondensible component, as when water vaporises into or condenses from a stream of air, the mass-transfer rate depends upon a mass-transfer coefficient on the vapour side of the interface, as well as upon the heat transfers from vapour to interface and from interface to liquid. The way in which these processes interact so as to determine the rate of mass-transfer is well known, and expressible by way of simple formulae; the only worry for the numerical analyst concerns the extent to which, if the mass-transfer rate is calculated explicitly with their aid, numerical instability may arise. There is space here only to indicate qualitatively how this possibility arises, and what should be done about it if it should become serious.

The mass-transfer rate <u>depends upon</u> three quantities which it <u>also influences</u>; they are:- the liquid-stream enthalpy; the vapour-stream enthalpy; and the vapour-stream composition. If the dependences and the influences are both strong, in the sense of the resulting fractional changes being large compared with the fractional changes which cause them, instabilities will occur.

By sufficient reduction of the time step, the condition can always be avoided; but this reduction may entail an unacceptable computer-time-expense penalty. Therefore, a Newton-Raphson technique must be employed, according to which:-

● the dependences of m upon the three variables are

expressed in a linearised form;

- the appropriately linearised form of the m expression is substituted into the finite-domain equation for each of the three variables before the equation is solved;

- if necessary, even more "implicitness" is built in by the simultaneous solution of the equations for all three variables.

(ii) Another situation, of less importance for nuclear engineering, arises when the mass-transfer rate is associated with a chemical reaction. In this case, the appropriate chemical-kinetic laws must be introduced; and, since these usually entail a strong temperature dependence, the interactions with the enthalpy-balance equations are again present.

Once again, the possibility of instability arises when the time-step is sufficiently large; and the general principle of the remedy is the same as has just been prescribed.

3.6 Comparisons between alternative successful formulations and procedures

As mentioned earlier, several authors have been successful in performing multi-phase-flow computations. It is therefore interesting to inquire as to whether their methods all fit into the theoretical framework just described; and, if so, in what respects they differ. Correct answers are not easy to give; but they will now be attempted.

It appears to be the case that nearly all the successful computations of multi-phase flows by numerical methods employ methods deriving from either Los Alamos Scientific Laboratory (LASL) or from the present author and his associates. Discussion will therefore be confined to these two streams of work.

Both streams fit into the framework described above. They employ the "staggered grid", the same conservation laws, "upwind" (ie "donor-cell") differencing, and the use of a Poisson equation for the pressure.

The differences lie in details. The LASL methods incline more towards explicitness (ie the finite-domain equations for velocity contain more beginning-of-interval values than the author would use), and towards lack of symmetry (ie greater prominence for one phase as compared with the other). The author's methods incline towards full implicitness (so that steady-state solutions can be obtained by setting the time step to infinity). and towards generality (so that the same formulations can be used for flows with chemical reaction, for example).

However, both groups are evidently experimenting, as is natural and proper; and each successive publication usually contains one or more modifications, introduced either so as to facilitate the solution of a particular problem, or recommended for general employment in the light of experience.

A detectable trend is for the LASL methods to increase in implicitness, as time goes on; and another is for the author's method to discard complications (for example, allowance for neighbouring r and R corrections as well as p and P corrections) which, although well-meant, proved more troublesome than helpful.

From the point of view of the on-looker, and potential user of the methods, the most important observations may be:-

- Although alternative routes towards the solution would be taken by adherents to the two streams, the solutions themselves (if the time-steps were small enough) would be the same.
- Differences in efficiency no doubt exist; but more is likely to depend upon the skill of the writer of the computer program than upon the intrinsic characteristics of the method.
- Possibly the LASL methods may show an advantage for rapid transients, and the author's methods for slow transients and the steady state. But, if ever such a difference were shown to be economically significant, the laggard would be able easily to make changes enabling him to catch up.
- Attention is better devoted to questions concerning the expressions for the inter-phase transfers of momentum, heat and mass, and especially concerning their realism and generality. However, it must be recognised that neither group of methods is any more bound intrinsically to the transfer relations which happen to have been used than to the velocities, pressures and concentrations which happen to have appeared in the published computations.

It does, however, need to be said that the expense of solving large numbers of finite-domain equations, which are needed if the representation of a three-dimensional multi-phase flow process is to be realistic, remains distressingly high. The ingenuity of the numerical analyst therefore needs to be steadily applied to the devising of efficient solution algorithms, and economical computer codes. Sponsors of code-development projects exert a great influence here, often by default: were they to be issuing contracts to competing organizations to solve the same equations in more economical ways, the users' computer costs would be reduced by orders of magnitude.

4. PHYSICAL INPUTS TO THE EQUATIONS

4.1 Turbulence models for multi-phase flows

(a) The problem

Equation (3.2-9) describes, in a compact form, how the volume fraction r in a cell is influenced by the various flow rates. When the flow is turbulent, it is useful to concentrate attention on the values of quantities which are averaged over times which are long compared with the durations of the turbulent fluctuations.

Let these average quantities be denoted by over-bars. Then equation (3.2-9) takes the form:

$$\overline{rg_0} = \overline{r_i}.\overline{g_i} + \overline{m} + [(\overline{rg_0}-\overline{r}\overline{g_0}) + (\overline{r_i.g_i}-\overline{r_i}.\overline{g_i})]$$

.(4.1-1)

Here, the quantities in the square brackets can be regarded as the consequences of the turbulent fluctuations of the g's being "correlated" with simultaneous fluctuations of the r's which they carry.

In single-phase turbulence-modelling practice, such turbulence fluxes are commonly expressed by way of effective exchange coefficients, multiplied by the gradients of the relevant scalar quantities. In accordance with this practice, equation (4.1-1) would be written as:

$$\overline{rg_0} = \overline{r_i}.\overline{g_i} + \overline{m} + Y_i.(\overline{r_i}-\overline{r})$$

,(4.1-2)

or:

$$\overline{r}(\overline{g_0}+Y_i) = \overline{r_i}.(\overline{g_i}+Y_i) + \overline{m}$$

.(4.1-3)

Here Y_i represents the product of an exchange coefficient, a cell area, and the reciprocal of an inter-node distance, and summation terms have been omitted, for compactness, as before.

The problem is: if equation (4.1-3) is regarded as defining the Y's, how are these quantities to be given actual values?

(b) Discussion

The concepts of turbulent mixing which have guided the thoughts of single-phase turbulence modellers derive from the early publications of Reynolds [38], Boussinesq [39] and Prandtl [40].

These authors have imagined:

- a surface in a fluid across which there are fluctuations of velocity, but no net flow of mass;
- a scalar variable, eg the mass fraction of some inert chemical species;
- a larger value of that variable prevailing on the one (richer) side of the surface than on the other (weaker); and
- a net transport rate of the quantity measured by the scalar variable, across the surface.

Their notion has then been that the fluid crossing from the richer to the weaker side carries more of this scalar quantity than the fluid passing in the other direction; the net transport rate is thus directed from "rich" to "poor", and is proportional to the difference of mass fraction.

Is this notion still applicable when two-phase flows are in question?

An immediate conceptual difficulty arises when it is fluctuations in g that are being looked at: there is no question of fluid being richer in light-phase material when it is going one way rather than another; it is 100% light-phase material at all times.

The difficulty is not simply a conceptual one: in practice, a turbulent two-phase flow in a pipe, for example, does not arrange itself so that the volume fractions of the two phases become uniform over the cross-section after the evening out of initial discontinuities. Instead, variations of r and R with radius persist indefinitely.

(c) Some implications for single-phase turbulence modelling

Pondering questions relating to turbulence in two-phase mixtures has led the present author to recognise that single-phase turbulent flows behave in many respects like two-phase ones. "Intermittency" for example, which is measured by experimentalist, and which is visibly manifest in the fragmented appearance of smoke plumes from chimneys, can be expressed mathematically in terms of "volume fractions" of turbulent and non-turbulent, or smoky and clear, intermingled gaseous phases: and the whole mathematical apparatus which has been developed for two-phase flows of the stream-water variety is proving to be applicable to its quantitative prediction.

There is no space to describe this development here; but papers which enable its course to be followed are references [55 to 61].

4.2   Flow-regime prediction

(a)   The problem

The considerations of section 4.1 are meaningful already for the very simplest of two-phase flows, for example, the transport of uniformly-sized spheres of solid in an incompressible liquid.

In nuclear-reactor thermal hydraulics, however, further problems arise from the fact that the phases may be interspersed in a variety of forms (droplets, bubbles, films), and with a wide distribution of sizes of the associated geometrical elements.

Obviously, even when all has been learned about turbulence modelling for two phases which are interspersed in prescribed manners, it will be necessary, in order to use the information, to have a scheme which will predict, for example:-

●   what proportion of the liquid is in droplet form;
●   what is the average size of the droplets, and (preferably) what is the droplet-size distribution;
●   what proportion of water is enclosing bubbles;
●   what is the average size (and, preferably, the size distribution) of these bubbles;
●   what is the proportion of liquid which forms a film on interspersed solid structure;
●   etc.

This need is, rather loosely, called the problem of "flow-regime prediction".   However, the name, as commonly interpreted, scarcely does justice to the magnitude of the problem; for it may be thought to imply that the correct selection of one of the popular qualitative labels ("bubble-flow", "annular-flow", "churn-flow", etc) will suffice.   It assuredly will not.

(b)   Discussion

Qualitative labels can be selected, for example by reference to one of the so-called "flow-regime maps" [42, 43, 44, 45]. These carry the implication that, more or less, the flow regime depends upon some such quantities as the mass flow rate (of both phases) per unit area, and the ratio of the flow rates of the two phases.

The primitive nature of these prescriptions can be deduced from the facts that the so-called determining factors cannot usually be expressed in dimensionless form, and that too few data are available which would permit correlations with, say, density ratio or Reynolds number even to be sought.

It is worth considering what path a more quantitative approach might take.   A minimum measure of the interspersions of the two phases is the interface area per unit volume, A''', say.

This is likely to increase when the fluid is strongly sheared (large bubbles or droplets will be ruptured, films blown off their supporting surfaces); to diminish when the fluid is allowed to settle (large bubbles will rise, swallowing smaller ones; droplets will impinge on surfaces and form films); and to be influenced by surface-tension forces (perhaps hindering changes in A''').

When turbulence models for single-phase flows were being invented, notions of what quantities to concentrate upon, and what equations might describe their evolution, were no more clear than those relating to A'''. Should one not follow the turbulence-modeller's example and boldly postulate equations such as:

$$\frac{DA'''}{Dt} = S_+ - S_- \qquad\qquad ,(4.2-1)$$

where S_+ is some source-term expression which increases with the amount of shearing (including interphase shearing) work which is being done on the two phases together, while S_- is a corresponding sink-term expression?

Of course, one will not get the expressions right at first; and, even when an appropriate form for them has been established, experimental investigations will be needed, for the establishment of the "constants" which they contain, and for testing the universality of their values. It is equally certain that A''' will not alone suffice to characterise a two-phase mixture; but some percipient and fortunate person may recognise that, together with perhaps two other mixture properties, it provides enough to be going on with. No scientific peak of this loftiness was ever scaled at the first assault.

What is necessary is for some bold spirit to begin, for others to provide him with the necessary funds (and freedom from excessive pressure for immediate results), and for the by-standers to refrain from discouraging him by mocking his fool-hardiness.

Turbulence models for single-phase flows, which are now regarded (too lightly, in the writer's view) as well-established, came into existence because the above three conditions were (just) fulfilled. Can not a positive encouragement be given for the new advance into two-phase-flow territory? It is greatly needed, both for turbulence modelling and for flow-regime prediction.

4.3   Inter-phase-transport relations

(a)   The problem

Equation (3.5-1) exemplifies how the velocity components of the two phases at a point on the numerical grid are bound together by some inter-phase friction relationship; and equations (3.5-5) and (3.5-6) indicate how corresponding relationships for calculation of the various terms which appear in the finite-domain equations referred to above, ie those which represent the momentum and enthalpy balances of the distinct phases? What formulae indeed are available?

(b)   Discussion

These questions will be answered only obliquely, by the statement that, in the author's view, the formulae which exist, and which are perforce used in steam-generator and other codes, are little more than hopeful guesses, based upon single-phase experience. Even that experience is of doubtful relevance; for who can point to reliable pressure-drop and heat-transfer formulae for oblique single-phase flow through tube banks?

The most important points to emphasize therefore, in a survey such as this, are: -

●   available formulae are of dubious value;
●   experimental research is needed, in which the influences of the relevant (dimensionless) variables are systematically explored;
●   the experiments should be paralleled by numerical computations of the two-phase flow behaviour in the inter-tube spaces, and in the films flowing over the tube surfaces.

Steam generators exhibit structural elements of less conventionality than tube banks, for example: - tube-support plates, flow-distributing perforated baffles, and steam-water separators. The reactor itself, and the interconnecting pipes, plena and valves, present an even greater variety of solid protrusions into spaces through which steam-water mixtures must flow.

Whenever these protrusions have dimensions which are smaller than those of the cells of the grid on which the numerical computation is being conducted, the problem presented by the steam-generator tube bank appears in a new form. The response of the prudent numerical modeller must therefore be the same, namely: -
●   use whatever generalisation of single-phase knowledge appears plausible (but use it with caution); and
●   seek to institute an experimental-cum-numerical investigation, on the scale of the protrusion itself, so as to provide in time a more rational substitute.

5.   SUMMARY OF CURRENT CAPABILITIES

5.1   Computational

As will already be apparent, the present author's view is that the computational problems of multi-phase flow, which appeared formidable at one time, have been resolved for practical purposes (just why the "proofs" regarding non-hyperbolicity proved to be so misleading can be left to the mathematicians to determine).

The central difficulty turned out to be that of determining an appropriate formula for the pressure correction (or corrections); but several suitable ones have been discovered; and, with their aid, it is proving just as easy to solve multi-phase-flow problems, whether the distinct phases have their own pressures or share a common one, as it is to solve single-phase ones.

No doubt, improvements can and will be made; but there is no need whatsoever for any further doubt about the availability of robust and general solution procedures.

5.2   Large-scale applications

The solution procedures have been applied to various one-dimensional flow phenomena [14, 15, 16, 17, 50, 51] and to multi-dimensional ones involving complex equipment [7, 8, 21-24, 47, 49]. The computer codes exist, and are becoming ever more widely used. However, because of the large component of guess-work which enters the computations, for reasons discussed in section 4, the predictions themselves can safely be used only as indicators of what might occur in practice.

Properly understood, this capability can be of great practical value; for computer codes can easily be operated over the whole range of possible values for the doubtful input quantities. It is therefore possible to draw reliable conclusions as to the reasonable limits of equipment behaviour, to rule out some possibilities with confidence, and cautiously to indicate others as being probable.

In naive hands, it must be emphasised, the large equipment-simulating computer codes can be highly dangerous. The sign of naivety which must be watched for is the uttering of sentences such as: "ATHOS predicts that the maximum slip ratio in the ..... steam generator is .....", and "TRAC has demonstrated that the maximum fuel-rod temperature in the ..... simulated accident will be .....". When all references to the particular flow-regime, inter-phase transport and turbulent-mixing presumptions are omitted, inadequate understanding of the vital role of these inputs is to be suspected.

5.3   Small-scale applications

What it is possible to predict for large structure-filled spaces can be predicted even more easily on the smaller scale of: -
● the spaces between tubes or rods in a bundle;
● the interior of a steam separator;
● the holes of a flow-distributing baffle.

Although doubts about flow regime, interfacial area and

turbulent mixing still remain, they are less severe because of the more precise specification of initial and boundary conditions that it is possible to provide; and a good chance exists of making correct predictions for at least major features of the flow, such as film-thickness distribution, separator efficiency, etc.

The assertion, it must be admitted, is a matter more of confidence than achievement at the present time; but the next few years should provide sufficient support.

6. SUMMARY OF CURRENT NEEDS

6.1 Computational

Once the Wright Brothers had demonstrated the practicability of powered flight, the rest of the world could set about developing more powerful, efficient and reliable airplanes. So it is also with computational procedures for multi-phase flow.

This notion, that now anyone can construct a multi-phase computer code, and can profit from the experience of (and so surpass) the pioneers, appears not to be widely recognised by funding agencies. There is therefore, in the writer's opinion, too little competition in the field.

Having paid a large sum to one organization to produce the first computer code for a particular kind of computation, funding agencies are understandably reluctant to admit that someone else can produce a second and much more efficient code, a year later, for one tenth of the cost; and this reluctance becomes rationalised in various ways. Yet second codes can be much better and cheaper than pioneering ones; and superiority factors of 10 are quite common.

Of course, novelty is not desirable for its own sake; and newcomers may promise more than they will find, as their experience grows, they can actually perform. Nevertheless, the present near-monopoly position of a few code-producing organizations is, in the author's view, neither necessary nor healthy.

6.2 Large-scale applications and validations

The steam generator and the reactor of the pressurised-water system are perhaps among the largest of the two-phase equipment items on the nuclear horizon; but they are not the largest (the cooling tower is that); and the turbine and the condenser also exhibit two-phase-flow features which are deserving of numerical modelling. Add to these the likelihood of two-phase flow in pumps and plena under accident conditions, and the number of potential applications is seen to be a very large one.

In summary, therefore, it can be stated that means of analysis which have been pioneered in connexion with the reactor and the steam generator are ready to be applied to other important items in the nuclear-power plant and beyond.

The question of validation of predictions made for large power-plant items, and indeed complete systems, requires some remarks. To test the predictions is of course necessary; but it is so expensive that very few tests can ever be made; as a result, "good agreement" between predictions and experiments may be a poor indicator of the general reliability of the computer model.

Much better value for money is given by research directed towards validating or improving the small-scale-process formulations on which the whole-equipment prediction rests. Such matters are discussed extensively in reference [51], and more briefly in reference [50].

6.3 Small-scale processes

The dominant influence which is exerted over the predictions of TRAC, ATHOS, PHOENICS, etc by the physical-process formulae which happen to be called upon has been repeatedly emphasised above.

The inadequate theoretical and empirical basis of these formulae has been given equal emphasis, as well as the possibility of remedying it by the institution and conduct of the appropriate research.

Here, therefore, there is no need to do more than state that it is the most pressing need of all. Until a large-scale and well-coordinated research program has got well under way, all predictions concerning the safety of pressurised-water nuclear plant, for example, must be accepted only with great caution.

Finally, it is not only routine research that is needed at this level: concept-forming innovation is needed. Men capable of making important door-opening innovations are understandably (perhaps fortunately) rare; but surely the nuclear industry can find a few in its hour of need? Perhaps all that is necessary is widespread recognition of the need, and willingness on the part of the community to entertain new possibilities.

7. ACKNOWLEGEMENTS

The author wishes to acknowledge the collaboration of colleagues and students at Imperial College, Purdue University, CHAM Limited, and CHAM of NA Inc, in the work which underlies much of the present paper. Their names appear appropriately in the list of references.

8. NOMENCLATURE

Symbol	Meaning
a	coefficient in a finite-domain equation
A	area of a cell wall
A'''	interfacial area, between phases, per unit volume
b	term in a finite-domain equation
c, C	phase-to-interface conductances per unit volume
d, D	densities of lighter and heavier phases
f, F	quantities defined by equations 3.3-33 and 3-35
g, G	product of phase density with either velocity times area or volume divided by time interval
g', G'	rate of change of g and G with pressure difference
g^*, G^*	values of g and G appropriate to p^* and P^*
h, H	specific enthalpy of lighter and heavier phases
k, K	turbulence kinetic energy of the two phases
l, L	turbulence length scale of the two phases
m, M	mass source of the two phases in a cell
p, P	pressures of the two phases
P^+	$P-p$
p^*, P^*, P^{+*}	guessed values of pressures
$p', P', P^{+'}$	corrections to guessed pressures
p''	p' for a neighboring cell
q, Q	heat sources of the two phases in a cell
r, R	volume fractions of the two phases
$R^{+'}$	rate of change of R with P^+
S_+, S_-	source and sink of interface area
δt	increment of time
u, U	velocities of the two phases in west-to-east direction
v, V	velocities of the two phases in south-to-north direction
w, W	velocities of the two phases in low-to-high direction
V	cell volume
α	alternative symbol for r
α, B, Y	constants in a quadratic equation
Y	exchange coefficient
λ	latent heat of phase change

Subscripts etc

N,S,E,W,H,L	north, south, east, west, high, low neighboring grid points or cells
n,s,e,w,h,l	the cell faces adjacent to the corresponding cells
P, P-	the cell (or grid point) at the current and previous times
o, i	outflow and inflow
sat	saturation (ie thermodynamic equilibrium)
*	guess, or appropriate to guess
'	correction, or appropriate to correction

+   extra, or appropriate to extra
−   time average

LIST OF REFERENCES

1.   SOO SL – Fluid Dynamics of Multi-phase Systems, Blaisdell, Waltham, 1967.

2.   WALLIS GB – One-dimensional Two-phase Flow, McGraw Hill, New York, 1969.

3.   GIDASPOW D – Modeling of Two-phase Flow Round Table Discussion (RT 1-2) Proc 5th Int Heat Transfer Conf Vol VII p 163, 1974.

4.   HARLOW F H – Comments at Round Table Discussion, (RT 1-2), Proc 5th Int Heat Transfer Conf, Vol VII, p 164, 1974.

5.   HARLOW F H & AMSDEN A A – Numerical calculation of Multi-phase Fluid Flow, J Comp Physics, Vol 17, No 1, pp 19-52, Jan 1975.

6.   SPALDING DB – The calculation of free-convection phenomena in gas-liquid mixtures, ICHMT Seminar, Dubrovnik, pub in Turbulent Buoyant Convection, Hemisphere, Washington, Eds N Afgan and D B Spalding pp 569-586, 1977.

7.   LILES DR & MAHAFFY JH – Modeling of Large Pressurised Water Reactors, in Basic Two-phase flow modelling in Reactor Safety and Performance, EPRI Conf, Tampa, Florida, March 2, 1979.

8.   SPALDING DB – Multi-phase flow prediction in power-system equipment and components, EPRI Workshop on Basic Two-phase Flow Modelling in Reactor Safety and Performance, Tampa, Florida, March 1979.

9.   CROWE CT & CHOI HN – Two-dimensional numerical model for steam-water flow in a sudden contraction, in: Two-phase Momentum Heat and Mass Transfer, Ed Durst F, Tsiklauri G, Afgan N, Hemisphere, Washington 1979.

10.   HARLOW FH & AMSDEN AA – Multi-fluid flow calculations at all Mach numbers, J Comp Physics, Vol 16, p1, 1974.

11.   RAMSHAW JD & TRAPP JA – A Numerical Treatment for Low-speed Homogeneous Two-phase Flow with Sharp Interfaces, J Comp Phys, Vol 21, pp 438-453, 1976.

40

12. MAXWELL TT & SPALDING DB – Numerical Modelling of Free-surface Flows, presented at Pressure Vessels and Piping Conf, San Francisco, Aug 12-15 1980.

13. SPALDING DB – Numerical computation of multi-phase fluid flow and heat transfer, in Recent Advances in Numerical Mechanics, Ed C Taylor, Pineridge Prtess, 1980.

14. SPALDING DB – One Dimensional Two-phase Flow with Unequal Velocities, Imperial College, London, MED Heat Transfer Section Report Ref HTS/79/10, 1979.

15. MOULT A, PRATAP V S & SPALDING DB – Calculation of unsteady one-dimensional Two-phase Flows, First Int Conf on PhysicoChemical Hydrodynamics, Advance Publications, Channel Islands, 1978.

16. BAGHDADI AHA, ROSTEN HI, SINGHAL AK, TATCHELL DG, SPALDING DB – Finite-Difference Predictions of Waves in Stratified Gas-Liquid Flows", Proc ICHMT 1978 Conf, Two-phase Momentum, Heat & Mass Transfer, Eds Durst, Afgan Tsiklauri, Hemisphere 1979, Vol 1, pp 471-483, 1979.

17. KUROSAKI Y & SPALDING DB – One-dimensional Unsteady Two-phase Flows with Interphase Slip: A numerical study, presented at 2nd Multi-phase Flow and Heat Transfer Symp – Workshop, Miami Beach, April 1979.

18. PUN WM, SPALDING DB, ROSTEN H, SVENSSON U – Calculation of 2D Steady Two-Phase Flows, Proc ICHMT 1978: Two-phase Momentum, Heat & Mass Transfer, Ed Durst, Tsiklauri, Afgan, Hemisphere, Vol 1, pp 461-470, 1979.

19. SPALDING DB – Mathematical modelling of fluid-mechanics, heat-transfer and mass-transfer processes, Imperial College, London, MED Heat Transfer Section Report, Ref HTS/80/1, 1980.

20 SPALDING DB – Numerical Computation of Multi-Phase Flows, (NCMPF), A Lecture Course at Purdue University, April 1979.

21. SINGHAL AK & SPALDING DB – Numerical Modelling of Two-Phase Flow in Steam Generators, 2nd Multi-phase Flow and Heat Transfer Symp, Miami Beach, April 1979.

22. SINGHAL AK, SPALDING DB & MARCHAND EO – Parametric Computations for Thermal-hydraulic Performance of a PWR Steam Generator, presented at ASME Nuclear Div Conf, Saratoga, New York, Oct 6-8, 1980.

23. SINGHAL AK, KEETON LW & SPALDING DB - Predictions of Thermal Hydraulics of a PWR Steam Generator by Using the Homogeneous and Two Slip-Flow Models, ASME National Heat Transfer Conf, Orlando, Florida, July 27-30, 1980.

24. MARCHAND EO, SINGHAL AK & SPALDING DB - Predictions of Operation Transients for a Steam Generator of a PWR Nuclear Power System, ASME Nuclear Division Conf, San Francisco, August 18-21, 1980.

25. ISHII M - Thermo-fluid Dynamic Theory in Two-Phase Flow, Reyrolles, Paris, 1975.

26. BANERJEE S & CHAN AMC - Models for Transient Two-phase Flows, in Basic Two-phase Flow Modeling in Reactor Safety and Performance, EPRI Conference, Tampa, Florida, March 2, 1979.

27. DELHAYE JM - Space-averaged Equations and Two-phase Modelling, in Two-phase Momentum, Heat and Mass Transfer, Eds Durst F, Tsiklauri GV and Afgan N H, Hemisphere, Washington, 1979.

28. LAUNDER BE & SPALDING DB - Mathematical Models of Turbulence, Academic Press, London & New York, 1972.

29. HARLOW F H & WELCH JE - Numerical calculation of Time-dependent Viscous Incompressible Flow of Fluid with Free Surface, Physics of Fluids, Vol 8, No 12, pp 2182 - 2189, 1965.

30. AMSDEN AA & HARLOW FH - The SMAC Method, Los Alamos Scientific Laboratory Report No LA-4370, 1970.

31. AMSDEN AA & HARLOW FH - A Simplified MAC Technique for Incompressible Fluid Flow Calculations, Los Alamos Report No LA-DC-11272, 1970.

32. HARLOW FH & AMSDEN AA - Numerical Calculation of Almost Incompressible Flow, J Comp Physics, Vol 3, No 1, 1968.

33. PATANKAR SV & SPALDING DB - A Calculation Procedure for Heat, Mass and Momentum Transfer in Three-dimensional Parabolic Flows, Int J of Heat and Mass Transfer, Vol 15, pp 1787-1806, Pergamon Press 1972.

34. PRATAP VS & SPALDING DB - Numerical Computations of the flow in Curved Ducts, Aero Quarterly, Vol 26, pp 219-228, Aug 1975.

42

35.   PATANKAR SV – Numerical Heat Transfer and Fluid Flow, Hemisphere Publishing Corporation, 1980.

36.   SPALDING DB – Mathematical Modelling of Fluid Mechanics, Heat Transfer and Mass Transfer Processes, a lecture course, Imperial College, London, MED Heat Transfer Section, Ref HTS/80/1, 1980.

37.   CARETTO LS, GOSMAN AD, PATANKAR SV & SPALDING DB – Two Calculation Procedures for Steady, Three-dimensional Flows with Recirculation, Proceedings of the 3rd Int Conf on Numerical Methods in Fluid Mechs, Vol II, pp 60–68, Pub. Springer-Verlag Heidelberg, Ed J Ehlers, K Hepp, H A Weidenmuller, 1973.

38.   REYNOLDS O – On the Extent and Action of the Heating Surface for Steam Boilers, Proc Manchester Lib Phil Soc, Vol 8, 1874.

39.   BOUSSINESQ J – Theorie de l'ecoulement tourbillant, Mem Acad, Sci, Vol 23, No 46, 1877.

40.   PRANDTL L – Bericht uber Untersuchungen zur ausgebildeten Turbulenz, Z angew Math Mech (ZAMM), Vol 5, No 2, pp 136–139.

41.   LAHEY RT & DREW DA – The Analysis of Phase Distribution in Fully-developed Two-phase Flows, 2nd Multi-phase Flow and Heat Transfer Symp, Miami Beach, April 1979.

42.   BENNETT AW ET AL – Flow Visualisation Studies of Boiling at High Pressure, Proc Inst Mech Eng, 1965–6, 180 (3c).

43.   SCOTT DS – Properties of co-current gas-liquid flow, Adv Chem Engg, Vol 4, pp 474–481, 1963.

44.   SCHICHT H H – Flow Patterns for an Adiabatic Two-phase Flow of Water and Air within a Horizontal Tube, Verfahrenstechnik 1969, Vol 3, No 4 pp 153–161.

45.   HEWITT G F & ROBERTS D N – Studies of Two-phase Flow Patterns by Simultaneous X-ray and flash photography, UKAEA Ref No AERE M2159, 1969.

46.   HULME G, PHELPS PJ, SPALDING DB & TATCHELL DG – Prediction of Steady, Three-dimensional Flow in Pressurized-water Steam Generators, In Boiler Dynamics and Control in Nuclear Power Stations, BNES London 1979.

47.   SPALDING DB – A General-Purpose Program for Multi-Dimensional One- and Two-Phase Flow, J Mathematics and Computers in Simulation North Holland, Vol XXIII, pp 267–276, 1981.

48. SPALDING DB – Developments in the IPSA Procedure for Numerical Computation of Multiphase–Flow Phenomena with Interphase Slip, Unequal Temperatures, etc, in Numerical Properties and Methodologies in Heat Transfer – Proceedings of the Second National Symposium, Ed. T-M Shih, Maryland, Chapter 6, pp 421-436, 1983.

49. SPALDING D B and MARKATOS NC – Computer Simulation of Multiphase Flows – a course of 12 Lectures, Notes of a course delivered at Imperial College in June 1983, CFDU Report CFD/83/4.

50. SPALDING DB – Computer simulation of Loss-of-coolant Accidents in Pressurized-water-reactor Nuclear Plants, Int Conf on Numerical Methods for Transient and Coupled Problems, Venice 9-13 July 1984.

51. KENNING DRB, SIMPSON HC and SPALDING DB – The reliability of the thermal-hydraulic calculations appearing in the Sizewell Pre-construction Safety Report, Report to the Nuclear Installations Inspectorate by the LOTS Sub-group, February 1984. Document submitted by NII to the Sizewell Public Enquiry.

52. HETSRONI G (Ed) – Handbook of Multi-Phase Systems, Published by Hemisphere, Washington, 1982.

53. BERGLES AE, COLLIER JG, DELHAYE JM, HEWITT GF, and MAYINGER G – Two-phase flow and heat transfer in the power and process industries, Hemisphere, Washington DC, 1981.

54. KAKAC S AND ISHII M (Eds) – Advances in Two-Phase Flow and Heat transfer, Vols 1 & 2, NATO Advanced Study Institute Series, Martinus Nijhoff, Boston, 1983.

55. SPALDING DB, – Computers, turbulence and two-phase flow, Imperial College CFDU report, ref CFD/82/3, January, 1982.

56. MA ASC, SPALDING DB AND SUN RLT – Application of "ESCIMO" to the turbulent hydrogen-air diffusion flame, Imperial College CFDU report, ref CFD/82/2, January 1982, later published in Proceedings 19th Symp (Int) on Combustion/The Combustion Institute, 1982/pp 393-402.

57. SPALDING DB – Chemical Reaction in Turbulent Fluids, Imperial College CFDU report, ref CFD/82/8, June 1982, later published in PCH Volume 4 No 4 pp 323-336, 1983.

58.   SPALDING DB – Towards a two-fluid model of turbulent combustion in gases, with special reference to the spark-ignition engine, in International Conference on Combustion in Engineering, Vol 1, Inst Mech Eng, 1983, paper C53/83.

59.   SPALDING DB – Turbulence modelling – a state-of-the-art review, Imperial College, CFDU report, ref CFD/83/3, July 1983.

60.   SPALDING DB – The two-fluid model of turbulence applied to combustion phenomena, AIAA 22nd Aerospace Sciences Meeting, Reno, Nevada, Paper No AIAA 84-0476, 1984.

CHAPTER 2

FINITE ELEMENT CALCULATIONS OF TEMPERATURE FIELDS IN DRY-STOR-
AGE CONTAINERS FILLED WITH SPENT NUCLEAR FUEL ELEMENTS

J.F.Stelzer

Kernforschungsanlage, KFA (Nuclear Research Centre),
5170 Jülich, FR Germany

SUMMARY

An interim solution for the storage problem of spent nuclear fuel is offered by gas-tight containers of cast iron. Inside of them the fuel can be stored for a random time. For the
different types of fuel elements different casks were developed. There was the necessity to know the temperature distributions in the casks a priori. This paper covers the way these
calculations were executed. The treated cask types are: one
for fuel of boiling water reactors (BWR), another for fuel of
pressurized water reactors (PWR) and a third one for fuel
spheres from the high temperature reactor (HTR). A cask for
BWR fuel holds 16 elements with 49 or 64 fuel rods each. In
the case of PWR fuel 4 elements with 236 rods each are contained. And with the HTR fuel one cask contains two cans with
1000 spheres of 6 cm diameter each. The cavities are flooded
with helium. It turned out to be advantageous to execute finite
element calculations in three steps. In the first step the
plane temperature field in the middle cross section is analyzed, where the highest temperatures are to be expected. This
calculation was done with respect to every single detail of
the arrangement according to the principle that the results
of a numerical calculation meet only than the reality when all
influencing physical details are respected. The nonlinear calculations take into account the natural convection in the helium gaps, the heat radiations and the temperature depency of
the physical properties. Together with the radiation the skew
effects according to Lambert's cosine law are examined as well
as the energy correspondence between fuel rods not directly
adjacent. Using the temperature distribution of the plane
exactly examined, averaged properties can be established which
are, in a second step, utilized to conceive a simplified axisymmetric finite element model. This model serves for an approximate spatial temperature field analysis. With this model
also the influence of the axial heat source distribution can

48

be examined which may be linear or cosine shaped or distributed
according to the burnt-down fuel. In a third step, the transient temperature development after fuel loading is calculated
with the axisymmetric model. The calculations were verified by
measurements later on. It turned out that the relative calculation error at the point of the highest temperature was approximately 1 %, thus showing the finite element method to be a
very reliable analytical tool.

1   INTRODUCTION

     For the transport and storage of burnt-up nuclear fuel a
thick-walled container of cast iron (spheroidal graphite iron)
[1] was developed. Figure 1 shows a cross section.

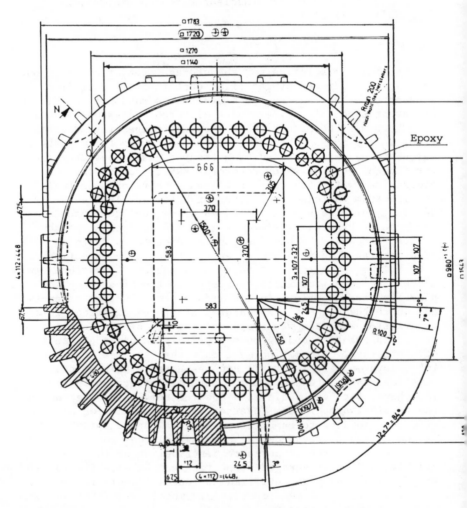

Fig.1 Cross section of a cast iron cask for spent fuel

The large central cavity holds 16 fuel elements in the case of
BWR fuel and 4 elements in the case of PWR fuel. With HTR fuel
two cans with 1000 sheres each are inserted. The BWR- and PWR
fuel elements are positioned in baskets of boron-alloyed steel.
Figure 2 shows a cross section of the PWR basket. In Figure 3
one PWR fuel element is sketched.

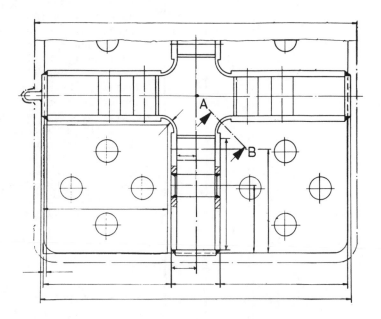

Fig.2   Basket cross section for PWR fuel elements

The fuel element types and dimensions are described in [2] and
[3]. The decay period of the fuel is usually one year. As well
the load set of the BWR fuel as that of the PWR fuel produce
about 19 kW. The different heat generation in the fuel elements
of both types causes the loading with different amounts of fuel
elements. In the HTR casks the heat generation is with 0.25 kW
comparatively low. The cavities in the casks are filled with
helium at atmospheric pressure. A large number of epoxy rods
are inserted into the cast iron wall to absorb delayed neutrons.
The container surface is equipped with fins. The steady state
and the transient temperature distribution in the containers
was of interest.

2   TWO-DIMENSIONAL MODELLING

It is well known from experience that the quality of analytical results depends on the consideration of all physical
details which determine the real system. In actual practice
this means the modelling of every fuel rod, every part of the
basket, distinction between rods with and without fuel (control

50

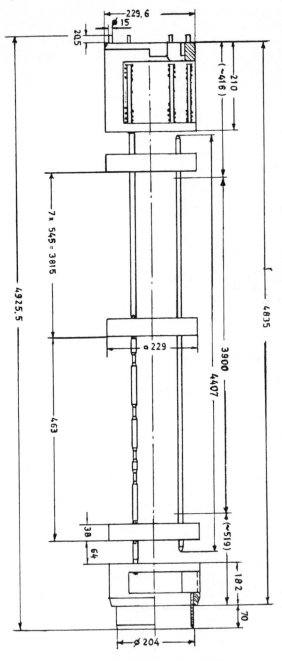

Fig.3   PWR fuel element (KWU)

rod), consideration
of the epoxy rods. It
further means the con-
sideration of nonlin-
ear influences: the
temperature depency of
thermal conductivity,
thermal radiation
across gas-filled gaps
also along skew paths,
and finally the impro-
vement of thermal con-
ductivity by natural
convection. As can
easily be seen, a com-
plete three-dimension-
al calculation with
these presuppositions
would be a very expen-
sive task. Therefore,
in the cases of the
storage of BWR and PWR
fuel, the very detailed
model was only created
for the hottest plane
at mid height. Fig.4
shows a quarter section
for the PWR fuel arran-
gement. Fig.5 displays
the appropriate finite
element mesh. The mo-
del consists of iso-
parametric 8-point ele-
ments with about 4500
nodal points. Of course,
instead of a quadrant
a 45 degree section
could have been suffi-
cient. However, there
would have been no
full symmetry because
of the epoxy rods. In
the case of the BWR
fuel element loading a
90 degree section had
to be modelled as well,
see Fig.6. This was
necessary since unsym-
metric loading took
place with fuel elements of different type. One type has 7 x 7
fuel rods, the other 8 x 8. The number of elements and node
points is approximately the same as with the first model. With
the HTR fuel loading a three-dimensional calculation could be

executed a priori because of the less complicated geometry. It will be treated in the 3-D model section.

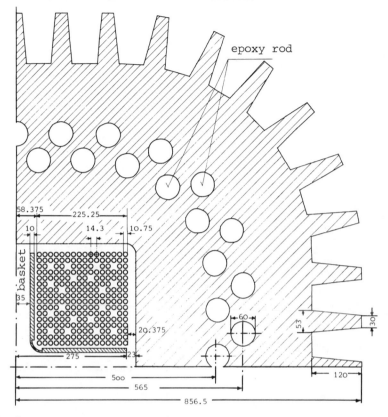

Fig.4 A quadrant of the plane arrangement for PWR fuel. The free positions belong to control rods.

Properties and boundary conditions. Every fuel rod cross section is modelled by one finite element. The assigned thermal conductivity is averaged according to the properties of cladding and fuel. Only the fuel rods are given the assignments of heat sources, but not the lead tubes of the control rods. The latter have a larger diameter, as can be seen in Fig.5. The helium in the gaps has a temperature dependent thermal conductivity which is expressed by a function given in [4]. The conductivity of the basket steel is also introduced by a temperature-dependent function from the same reference. The thermal conductivity for the cast iron was taken as a constant value because of its relatively small temperature increase. It is the value for ferritic spheroidal graphite iron. In connection with natural convection the dynamic viscosity and the density of the helium are also of interest. Their temperature depency was again regarded by using physical property algorithms from reference [4]. For thermal radiation, the emission coefficients of the radiating surfaces must be introduced. Finally, a suitable heat transfer coefficient is necessary for the outer surface. A value of

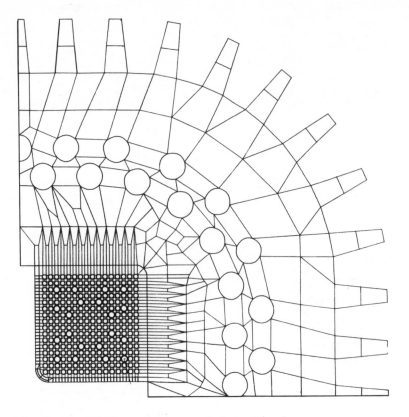

Fig.5 The finite element model to Fig.4

10 W/(m^{2}K) is adequate for the upright cask position.

The iteration procedure. The consideration of all nonlinearities is achieved by a direct iteration procedure, as described e.g. by Owen and Hinton [5]. Fig.7 shows how the temperature swings to its final value at the point of the highest temperature. 10 steps are necessary.

Convection in the gas-filled gaps. Besides the heat transfer by conduction and radiation across gas-filled gaps additional heat transport by natural convection is possible. This effect will appear when the gas begins to circulate as a result of density differences caused by temperature differences. It will occur when the product of Grashof- and Prandtl number exceeds 1700. Natural convection increases with the gas gap width. Helium has a low tendency to natural convection due to its high thermal conductivity. In our calculations the improvement factor due to convection is computed for every gas element by using the formula of Niemann[6]

$$\frac{K_f}{K} = 1 + \frac{m(GrPr)^r}{(GrPr)+n} \tag{1}$$

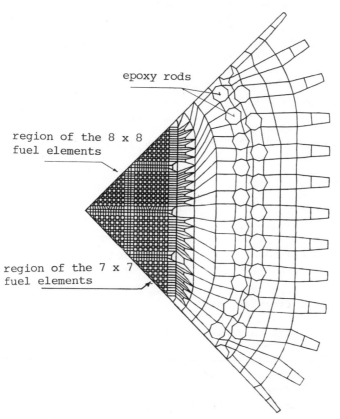

epoxy rods

region of the 8 x 8
fuel elements

region of the 7 x 7
fuel elements

Fig.6 The mesh for the BWR fuel element arrangement

with K_f being a fictitious heat conductivity including the improvement by convection and m = 0.0236, n = 10 100, r = 1.393 for a vertical slot. This K_f is used instead of the common local heat conductivity.

Thermal radiation heat transfer. The heat radiation from the outmost positioned fuel rods to the inner wall of the iron cask partly takes a straight course, see Fig.8. However, there is also heat transport along skew angles. The straight heat radiation is covered by the relationships given e.g. in [7]. They treat the radiation as an additional term of conductivity starting from the equation

$$C(T_1^4 - T_2^4) = \frac{K}{s} (T_1 - T_2) \qquad (2)$$

with $C = \sigma/(1/\varepsilon_1 + 1/\varepsilon_2 - 1)$ as the radiation coefficient, K as a term considering the radiation as a conduction phenomenon, s as the element width in the radiation direction. If the gas gap through which radiation takes place is modelled by a row of finite elements, then s is the whole gap width which must be introduced in every radiating element of this row. The radiation term follows as

$$K = sC(T_1^3 + T_1^2 T_2 + T_1 T_2^2 + T_2^3) \qquad (3)$$

54

This term is valid for the
straight radiation path of 7.6
mm width in Fig.8. The skew angle region is subdivided into
two areas from 0 to 45 degrees
and from 45 to 68 degrees. The
radiation is proportional to the
cosine of the angle of incidence.
For both regions an averaged
cosine is formed

$$\bar{\alpha} = \frac{1}{\Delta\alpha} \int_{\alpha_1}^{\alpha_2} \cos(\alpha) \, d\alpha. \qquad (4)$$

The sum of the appropriate numbers of both regions gives the
factor which has to be multiplied on the expression (3).

There is of course also
radiation heat transfer inside
the rod bundles, see Fig.9.The
radiation does not only pass
here along orthogonal channels
either as is given a priori by
the finite element net. There
are additional view areas to
elements farther away. The pro-

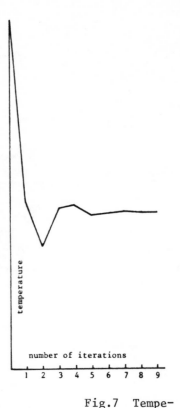

Fig.7 Tempe-
rature develop-
ment vs. number
of iterations

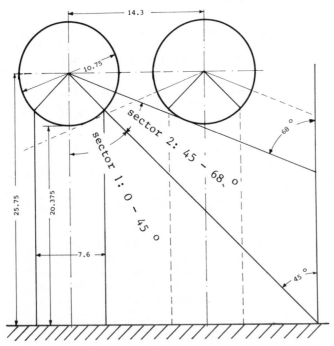

Fig.8 (left)
Consideration
of skew radiat-
ion

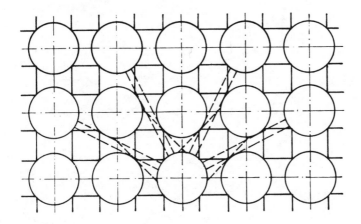

Fig.9   Paths of radiation in the rod bundle

gram deals primarily with the radiation between two adjacent
rod surfaces in the horizontal and vertical directions, respectively, see Fig.9. There are four radiation channels, 90 degrees each, giving a sum of 360 degress. Radiation correspondence to the rods in the next row farther away exists along
eight viewing strips, each of them being 21 degrees broad.This
gives another 168 degrees. Finally, four radiation channels
with 106 degrees width each exist to those neighbouring rods
which are positioned diagonally. They cover 424 degrees. The
sum of all these portions related to the primary part gives
the factor by which the radiation term must be multiplied.These
manipulations approximately cover the whole radiation heat
transfer. It should be mentioned that the radiation term K of
eq.(3) which is increased by the described manipulations is
simply added to the conductivity.

3   RESULTS OF THE TWO-DIMENSIONAL CALCULATIONS

A survey of the results is displayed in Fig.10. The gas
gap of 20 mm width is located between the contour lines no.5
(relative temperature 16.1) and no.6 (relative temperature
39.3). It causes a steep descent in the temperature profile.
Across the iron wall the temperature ascent is small. The epoxy
rods remarkably influence the temperature distribution. Another
result representation, which can be easily comprehended, is the
mountain-like apsect. Here, the mesh serves as a ground plan,
over the projection of which the appropriate temperature is
drawn at every mesh point. For the PWR fuel element arrangement a picture is generated as can be seen in Fig.11. The relationships in the case of the BWR array are shown in Fig.12.
The influence of the separating basket sheet can be seen
which form paths with good heat transport properties.
The partition of the heat transfer across the gas gaps
into the different transport mechanisms may be of interest.

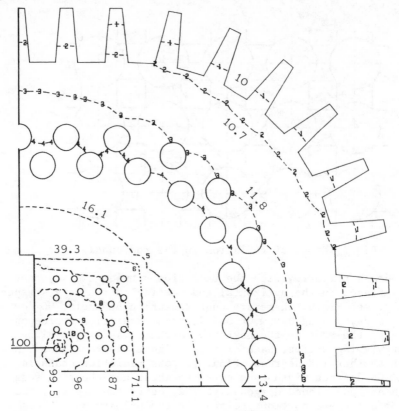

Fig.10 Contour lines plot of the results

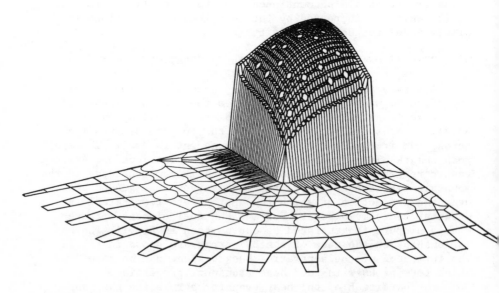

Fig.11 Result display in the mountain-like aspect

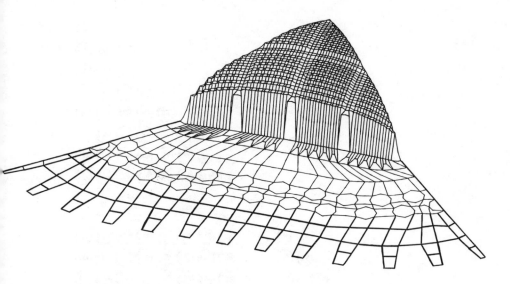

Fig.12   Temperature distribution in the BWR cask

In the wide gap between fuel elements and inner cask wall the relationships differ, dependent on the location, from a conductive part of 46.8 % to 59 %. The natural convection amounts to only 0.5 %. The remainder is radiation with 52.7 to 40.5 %. No convection takes place between the fuel rods. Here, the radiation plays de dominant role with approximately 77 %. The rest is conduction.

Figs.13 and 14 show digital results in normalized temperatures. After loading a cask with fuel elements equipped with thermocouples the temperatures at some characteristic locations were measured. These temperatures are specifiec at the appropriate locations. Since the temperatures are taken relatively, the differences between the numbers are the deviations expressed as percentages. E.g. in the hottest point of the calculation, relative temperature 100, the measured relative temperature is 101.1. This means a relative error of 1.1 %.

4   A SIMPLIFIED THREE-DIMENSIONAL MODEL

In the plane temperature field calculation the axial heat flux was neclected. A 3-D calculation seemed to be necessary to study this influence and to learn something about the temperature distribution at locations other than in the midplane. For this purpose, a simple axisymmetric model is established, where the complicated geometric relationships of the real arrangement are transferred as far as possible. The same is done with the inherent physical properties, viz. the radiation heat transfer across cavities and conduction across inhomogeneous regions. The first simplification consists in bringing the

58

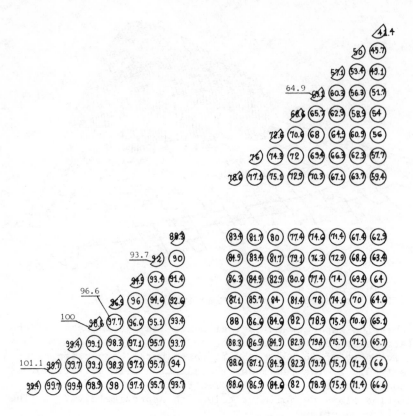

Fig.13 Relative temperatures in the region of the 8 x 8
rod arrangement, BWR. Measured temperatures are assigned

cask cross section into a circular shape, see Fig.15. Fig.16
shows the complete axisymmetric model. In the fuel element region there is an upper and a lower zone (15-degree hatching)
where no heat sources are present, and a 3900 mm long area
with internal heating. The fuel element feet and the lower part
of the basket are simulated by a rod (vertical hatching) of
equivalent conductance. The vertical gap is fully reproduced.

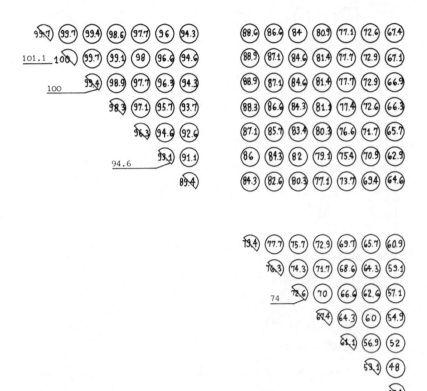

Fig.14 Relative temperatures in the area of the 7 x 7
rod arrangement

Physical properties. In a first step the heat sources ought to
be homogenized. This is done by dividing the whole generated
heat by the representative volume. The radial heat conductivity
in the fule region is calculated by using the equation which
combines the heat sources $\dot{q}$ (e.g. in W/cm^3), the temperatures,
the radius and the heat conductivity K, see ref. [8]

$$T(r) = T(r=0) - \dot{q}/(4K) \cdot r^2 . \qquad (5)$$

For determining the axial thermal conductivity a mixed value
is developed covering the different parallel heat flux paths

$$K_{ax} = a_1K_1 + a_2K_2 .. + a_nK_n , \; \Sigma a_i = 1$$

where a_i are the cross section shares of the different stuffs
(fuel rods, control rods, basket walls, helium) and K_i the individual heat conductivities. A special treatment is required
for the gas filled annulus. Here, the temperature depence of
the helium and heat radiation are taken into account. Moreover,
a special adjustment was necessary to obtain the same temperature jump across the gas gap as with the plane model. This was

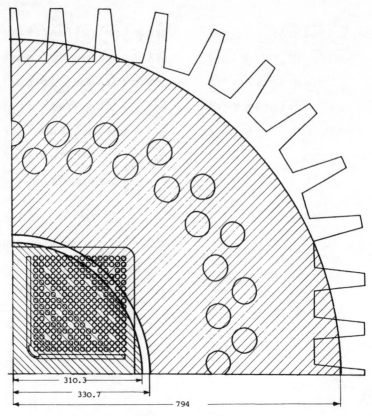

Fig.15 Transforming the original cross section into an equivalent circular shape

done iteratively with a simple model of three axisymmetric elements, see Fig.17. It turned out that it is advantageous to multiply a factor on that term that covers the heat radiation. Furthermore, the conductivity in the cast iron wall had to be adjusted in the radial direction using the well known formula

$$K_r = \dot{q}A \ln \left(\frac{r_o}{r_i}\right) / (2\pi(T_i - T_o)), \qquad (6)$$

with $\dot{q}A$ being the heat generation per length (W/cm), A the cross section of the homogenized fuel region, r_o and r_i the outer and inner radius, T_i and T_o the respective temperatures. Finally, the heat transfer coefficient on the cask surface must be adjusted to compensate for the reduction of the surface by omitting the fins. The adjusted heat transfer coefficient α can be gained from the expression

$$\dot{q} \, r_{fuel}^2 \, \pi = 2\pi r_o \, \alpha \, \Delta T \qquad (7)$$

with $\dot{q}$ being the specific heat generation in the fuel region,

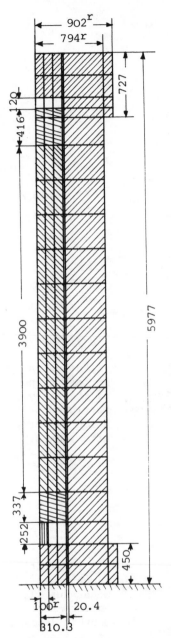

Fig.16 The axisymmetric
model

r_{fuel} its confining radius, ΔT the temperature difference on the outside from the ambient temperature to the surface temperature which is taken from the 2-D calculation. The substitute value is about three times that of the plane model. This substitute heat transfer coefficient is only applied in the finned area. Now, after having established the model and the adjusted properties, the analysis can begin.

Calculation and results. The neutron flux in the reactor is distributed along the vertical axis according to a cosine function with the maximum in the middle, at least in a first approximation. The burn-up is adequate and so are the heat sources in the fuel rods. The heat which is set free is about 15 % higher in the middle than at the ends. Two calculations were executed to assess this influence: Once with uniform heat distribution and a second time with the cosine distribution. The results obtained are shown in Fig.18. It is evident that the ridge of the temperature mountain with cosine heat distribution is more bent. The maximum temperature is, however, with cosine distribution only 3 % higher than with uniform heat source distribution. For this nonlinear calculations 11 iterations were necessary to receive convergence with a mean error of less than 0.1 %.

The knowledge of the spatial temperature field can be used to assess the relative error which appears in the 2-D calculation by neglecting the axial heat transport. Defining the error by

$$|\varepsilon| = (\Delta T_{2D} - \Delta T_{3D})/(\Delta T_{2D}) \qquad (8)$$

where ΔT is the temperature difference between the maximum temperature and the ambient temperature and the subscripts denote the 2-D- or 3-D case. An error of about 1 % results. This relationship can be made visible by plotting the local heat fluxes at the Gauss

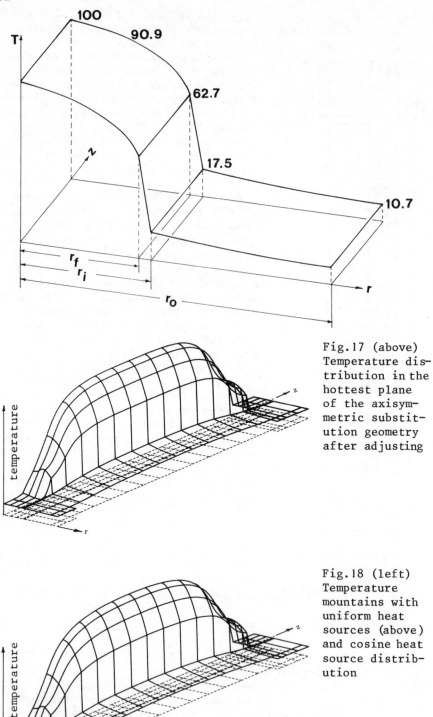

Fig.17 (above)
Temperature dis-
tribution in the
hottest plane
of the axisym-
metric substit-
ution geometry
after adjusting

Fig.18 (left)
Temperature
mountains with
uniform heat
sources (above)
and cosine heat
source distrib-
ution

positions, as is done in Fig.19, according to a proposal in ref. [9]. In midheight the heat flux arrows indicate horizontally outwards.

Of course, there is no problem to consider an axial heat source distribution which results from rather exact calculations with the Origene Code. Fig.20 displays such a distribution, and Fig.21 shows its approximation by a polynomial of fourth degree. The appropriate temperature field, this time shown in the vertical projection, can be seen in Fig.22. The temperature maximum is shifted to the lower third part. Fig.23 exhibits the pattern of the isothermal lines.

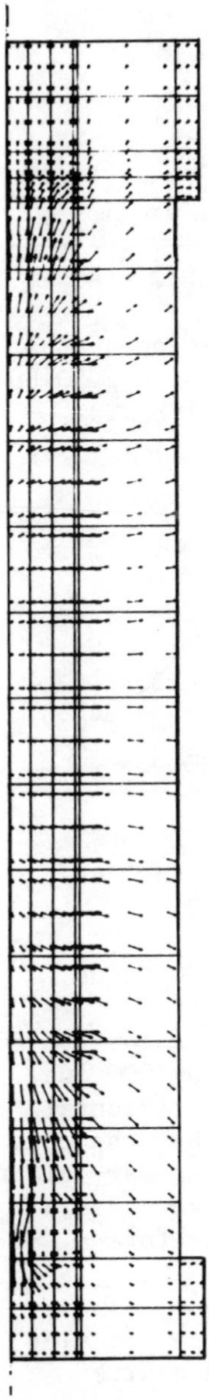

5 THE TRANSIENT TEMPERATURE FIELD

The axisymmetric model is used to examine the time-dependent behaviour of the temperatures. The nonlinearities in the transient case are treated according to chapter 21.9 in Zienkiewicz's book [10]. At the beginning, when the fuel elements are loaded into the cask, an uniform temperature of 27 °C is assumed. After about 4 hours half of the whole temperature difference between the cold and the steady-state final temperature is exceeded. In the time-marching scheme a time stepwidth of 900 s was used. Fig.24 shows the instantaneous temperatures at six selected points of time. In Fig.25 the time dependent development of some node temperatures in the hottest plane is plotted.

Fig.19 Local heat fluxes

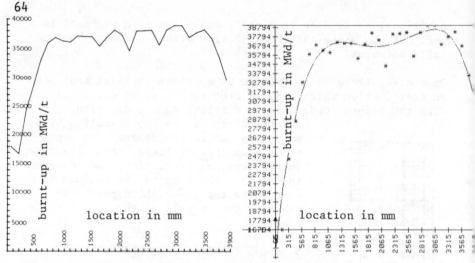

Fig.20 Location dependent
burnt-up in a certain fuel
element

Fig.21 Approximation of the
burnt-up pattern of Fig.20 by
a fourth order polynomial

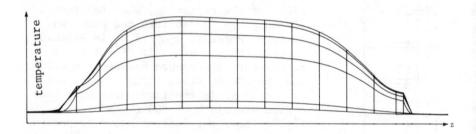

Fig.22 Vertical projection of the temperatures along the central axis with a source pattern according to Fig.21

6   THREE-DIMENSIONAL TEMPERATURE FIELD CALCULATIONS FOR
     CASKS WITH HTR FUEL

     The cast iron cask holds two cans filled with 1000 fuel
element spheres each. The cask is stored in the horinzontal
position as shown in Fig.26. The consequence is that the cans
sit excentrically in the cavity because of the necessary clear-
ance between can and wall. The appropriate influence should
be respected in the temperature calculations. Therefore a
three-dimensional calculation seemed necessary.

The dimensions and properties of the model and also the element discretisation can be seen in Fig.27. Only a quarter
section of the whole arrangement is modelled because of symmet-
ries. The pile of spheres is treated as homogeneous matter.
For the heat conductivity a suited value was calculated according to the proposals in ref. [11].

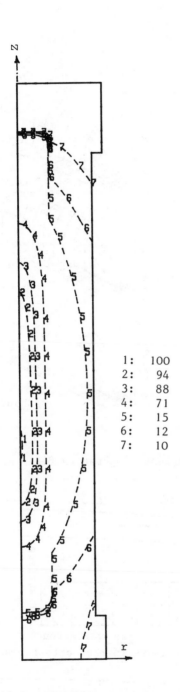

1:	100
2:	94
3:	88
4:	71
5:	15
6:	12
7:	10

Fig.23 Results in the display
of isothermal lines

The can is positioned excent-
ridally: the gas gap on the
left hand side is wide in con-
trast to that on the right
hand side. A survey of the results of the temperature field
calculation gives Fig.28. The
region of the highest temperature is visible, and also the
steep temperature gradient in
the wide gas gap.

7   ACKNOWLEDGEMENTS

This paper is based on
cooperation between DWK, Deut-
sche Gesellschaft für Wieder-
aufarbeitung von Kernbrenn-
stoffen (German company for
the recycling of nuclear fuel)
in Hanover, and GNS, Gesellschaft for Nuclear Service in
Essen, PROFEM GmbH in Aachen
and KFA, Kernforschungsanlage
(nuclear research centre) in
Jülich. Contributions were
made by K.Einfeld, J.Fleisch,
A.Lührmann and H.Spilker from
the DWK, by D.Methling from
the GNS, by J.Wimmer from the
PROFEM, and by E.Graudus and
A.Sievers from KFA. The 2-D
calculations were executed on
a IBM 3033 with the program
FEABL2 which is an advanced
version of the FEABL package
[12]. The element library and
the nonlinear procedures were
added by A.Sievers, J.F.Stelzer
and R.Welzel, KFA. The pre- and
postprocessing, especially all
graphics, and the calculations
with the axisymmetric model
were executed on a HP9845B
desktop computer with the FEM-
FAM software (in Basic language)
created by K.Groth, University
of Bonn, and J.F.Stelzer.

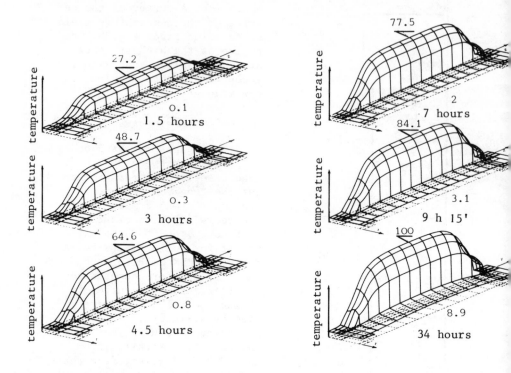

Fig.24 The development of the transient temperature field at six selected moments

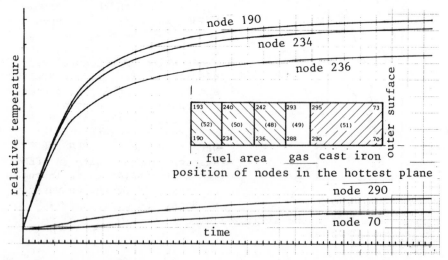

Fig.25 Time dependent temperature development at certain points

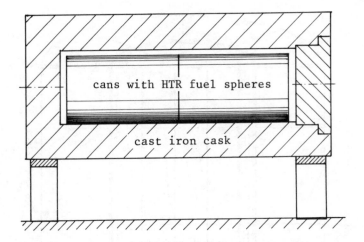

Fig.26 The storage arrangement of spent HTR fuel

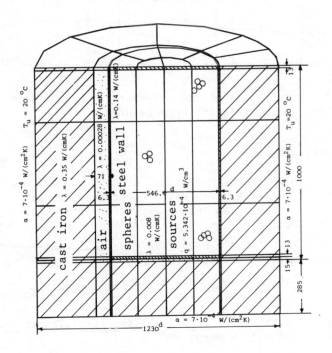

Fig.27 The model for the temperature field calculations of casks filled with HTR fuel

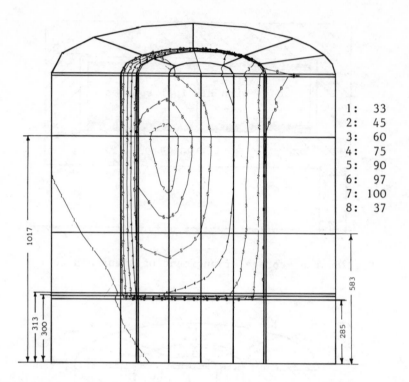

1:	33
2:	45
3:	60
4:	75
5:	90
6:	97
7:	100
8:	37

Fig.28 Result plot of the temperature calculations for
HTR fuel casks, relative temperatures

8 REFERENCES

1. DIERKES,P., K.JANBERG, H.BAATZ, G.WEINHOLD - Trans-
    port casks help solve spent fuel interim storage
    problems, Nuclear Engineering International,
    Oct.1980

2. KLUSMANN,A., H.VÖLCKER - Nuclear reactor fuel ele-
    ments, Karl Thiemig, Munich, 1969 (in German)

3. OLDEKOP,W. - Pressurized water reactors for power
    plants, Karl Thiemig, Munich, 1974 (in German)

4. STELZER,J.F. - Physical property algorithms, Karl
    Thiemig, Munich, 1984

5. OWEN,D.R.J., E.HINTON - Finite elements in plastici-
    ty, West Cross, Swansea, Pineridge Press, 1980

6. NIEMANN,H. - Heat transfer by natural convection in
    slots, Gesundheits-Ingenieur 69, 1948, 224-228
    (in German)

7. ROLPH,W.D., K.J.BATHE - An efficient algorithm for analysis of nonlinear heat transfer with phase changes, Int.Journ.Num.Meth.Engn., vol.18, 1982, 119-134

8. STELZER,J.F. - Heat transfer and fluid flow, Karl Thiemig, Munich, 1971 (in German)

9. STELZER,J.F. - Heat flux display and heat transmission coefficient calculation with the finite element method, Heat Transfer 1982, ed. U.Grigull, E.Hahne, K.Stephan, J.Straub, Hemisphere Publ.Corp., Washington, New York, London

10. ZIENKIEWICZ,O.C. - The finite element method, 3rd. ed., McGraw-Hill, London, 1977

11. LEYERS,H.J. - Heat conductivity in piles of loose spheres in stagnating media, Chemie-Ing.Technik, 1972, 1109-1115 (in German)

12. ORRINGER,O., SUSAN FRENCH - FEABL Finite element analysis basic library, AFOSR TR, ASRLTR 162-3, Mass.Inst.Techn., Cambridge, Mass., 1972

CHAPTER 3

EVALUATION OF THERMAL AND STRESS FIELDS IN
A COMPOSITE AXISYMMETRIC BODY WITH A NONLINEAR
SURFACE RADIATION BOUNDARY CONDITION

L. D. Wills and H. Wolf

ME Dept., Univ. of North Dakota, Grand Forks, N.D. 58201 USA
ME Dept., Univ. of Arkansas, Fayetteville, AR 72701 USA

1 SUMMARY

A general purpose, nonlinear, axisymmetric, finite element code (AXIFE) was constructed on the basis of the Poisson and Navier equilibrium equations; the code utilizes an algorithm that automatically implements a radiation heat transfer boundary condition on the surface of axisymmetric enclosures. The program AXIFE was successfully utilized to determine temperature fields, and subsequently stress fields, in an axisymmetric composite nuclear fuel rod whose centervoid surfaces were considered non-adiabatic and subject to surface radiation. Thermal results showed that radiation heat transfer in the centervoid lowers the maximum fuel temperature and shifts it toward the cladding; stress results severely limit rod power for performance with minimum cracking. The method, incorporating the radiation algorithm, requires the solution to a set of nonlinear equations and required three to ten times the computational time for axisymmetric enclosures with no radiation (adiabatic) depending on the accuracy of the initial temperature estimates. Results were generated for an axisymmetric body representing a section of a nuclear fuel rod; the code, however, has a much broader range of application. The fuel rod was modeled with twenty-one eight node, isoparametric, torroidal elements with quadratic serendipity interpolating functions. Specific temperature and stress results for the fuel rod are presented for 6 and 18 kw/ft (19.7 and 59 kw/m) rod power to illustrate the application of the radiation algorithm.

2 INTRODUCTION

Perhaps the most compelling feature of the finite element method in transport analysis is its adaptability to a wide variety of boundary conditions with little or no modification to the basic algorithm; the combined conduction-radiation problem discussed in this chapter is such an example. Several works have been reported in the literature [1], [2], that describe finite element codes utilizing the radiation heat transfer boundary condition, however, the main thrust of these works was based on a linearized model. In a physical situation where large temperature differences exist (and, consequently, thermal radiation becomes a significant portion of the energy transport mechanism) the use of such a linearized model yields appreciable error. Further most codes which treat radiation heat transfer require the manual calculation and input of configuration factors, an aspect which in itself constitutes a major portion of the analysts' effort.

A nuclear fuel rod is an important example of a composite axisymmetric structure that can be analyzed by the finite element method as described herein. In particular the surface temperatures of the centervoid (details of centervoid formation are discussed by Krudener [3]) are such that the temperature field is significantly affected by surface thermal radiation.

Current reactors use cylindrical fuel pellets in their fuel rods. Krudener and Wolf [4] have shown that a fuel module with preformed centervoids and separated with molybdeum washers (which serve as an auxillary heat transport path as well as a means of preventing fuel redistribution) permitted considerably higher power dissipation than annular fuel pellets at the same maximum fuel temperature. The analysis reported in Reference [4] have shown that a fuel module with preformed centervoids and separated with molybdeum washers (which serve as an auxiliary heat transport path as well as a means of preventing fuel redistribution) permitted considerably higher power dissipation than annular fuel pellets at the same maximum fuel temperature. The analysis reported in Reference [4] assumed an adiabatic centervoid. The results showed about 1500 degrees Rankine difference between the hottest and coolest surface of the centervoid; large enough for radiation to be significant. The power levels shown to be thermally possible brought into question whether the fuel pellet could survive the associated thermal stresses thereby providing the impetus for the creation of a more general finite element code; one that is applicable to the axisymmetric geometry and that could include the surface radiation boundary condition in the delineation of the thermal and stress fields.

Since the geometry of the void was subject to modification, thus requiring different sets of configuration factors, it was deemed advisable to formulate a method which would:
   a) handle the radiation boundary condition in its full nonlinear form,
   b) be adaptable to a wide variety of axisymmetric geometries, and
   c) require as little manual effort as possible.
By including in the code the required nonlinearity of the radiation boundary condition, the additional nonlinearities associated with thermal problems, such as temperature dependent

material and surface properties, could also be implemented with minimal increase in computation time.

3 PHYSICAL GEOMETRY

The example investigated with code AXIFE is shown in Figure 1. The entire fuel rod assembly is shown on the left of

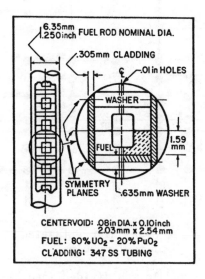

Fig. 1. Fuel module geometry.

the figure and a detailed enlargement of a module is shown on the right. The fuel module comprises a high conducting washer and two cup shaped fuel pellets so assembled that the insides of the cups constitute a preformed centervoid. Because of symmetry only one quadrant of the detailed enlargement (shaded areas indicated in Figure 1) was considered for analysis. Pertinent dimensions are shown on the figure and specific details are given by Wills [5] who made two modifications in the geometry that was described in References [6] and [4]. First, a 0.010 inch hole was included on the centerline of the fuel pellet and the molybdenum washer to avoid mathematical complications at r=0 and to facilitate fission gas migration into the upper plenum of the rod assembly; and second, a 0.005 inch radius fillet was added at the corner of the centervoid to more accurately reflect manufacturing methods.

Several fuel modules are assembled in a stainless steel tube 0.250 inches outside diameter with 12 thousandths inch wall thickness to comprise the fuel rod; several hundred rods then make up a fuel sub-assembly to be used in a reactor core.

Figure 2 shows in detail the discretization of the fuel module quadrant analyzed together with the salient dimensions. As a compromise between accuracy and computational ease, the

74

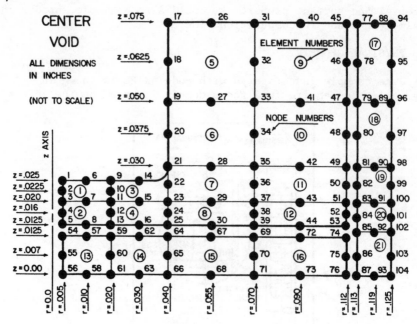

Fig. 2. Nodal discretization
of fuel module

quadrant of the fuel module was divided into 21 eight-node,
quadrilateral, torroidal elements; elements 1 through 12 in the
fuel, elements 17 through 21 in the cladding, and elements 13
through 16 in the washer to yield a sum total of 104 nodes.

Within each element a local coordinate system (ξ,η) was
defined such that $-1<\xi<1$ and $-1<\eta<1$. Also, a set of interpolating functions, N_i, were defined [5] at each node in terms of
the local coordinate system (ξ,η) as recommended by Zienkiewicz
[7] of the form

$$N_i = A(a_1 + a_2\xi + a_3\eta + a_4\xi\eta + a_5\xi^2 + a_6\eta^2 + a_7\eta\xi^2 + a_8\xi\eta^2)$$

$$a_i = +1, -1, 0 \quad (i=1,8)$$

(1)

where A is either 1/4 or 1/2 depending on the node number. From
these interpolating functions, the derivatives $\partial N_i/\partial\xi$ and
$\partial N_i/\partial\eta$ can be formed for later use in the governing equations
which describe the field variable. The choice of domain discretization and interpolating functions results in eight node
quadrilaterial, isoparametric, toroidal elements with quadratic
serendipity (Zienkiewicz [7]) interpolating functions. We now
direct our attention to the formulation of the governing equations which describe the behavior of the field variable over
the domain described above.

4 THE FINITE ELEMENT METHOD

In this work, the finite element method is applied to Poisson's equation delineating the thermal field and to the Navier equilibrium equation governing the stress field through use of Galerkin's method.

4.1 The Thermal Element Equations

Temperature variation in the three-region composite nuclear fuel module is governed by the elliptic partial differential equation, known as Poisson's equation, describing the steady state behavior of an isotropic domain in cylindrical coordinates for the axisymmetric case;

$$\partial[kr(\partial t/\partial r)]/\partial r + \partial[kr(\partial t/\partial z)]/\partial z + r\dot{q} = 0 \tag{2}$$

where t is temperature, r and z are radial and axial coordinates, $\dot{q}$ is unit volume generation rate, and k is thermal conductivity.

To apply Galerkin's method let the differential equation be $L(\phi)-f=0$ to which we desire the approximate polynomial solution $\tilde{\phi}$. The substitution of $\tilde{\phi}$ into the equation results in a residual $R=L(\tilde{\phi})-f$ rather than zero. Letting $\tilde{\phi}=N_i\phi_i$ (N_i are the interpolating functions of the previous section and ϕ_i are the values of the unknown field variable) it can be shown that when

$$\int_D [L(\Sigma N_i\phi_i) - f]N_i dD = 0 \tag{3}$$

for each of the points i in domain D that the residual R will tend toward zero when the set of unknown parameters ϕ_i, determined by the resultant equation set is substituted back into the assumed form of $\tilde{\phi}$. Applying Galerkin's technique to equation (2) results in a volume integral that can immediately be integrated over the circumferential direction θ thereby giving (after division by 2π) the result

$$\int_A kr^2N_i[(\partial^2t/\partial r^2) + (\partial^2t/\partial z^2)]drdz +$$

$$\int_A rN_i[k(\partial t/\partial r)+r\dot{q}]drdz = 0 \tag{4}$$

To proceed requires application of integration by parts in two dimensions to the first term of equation (4) as follows

$$\int_A u(\partial v/\partial x)dA = \int_S uvn_x ds - \int_A v(\partial u/\partial x)dA \tag{5}$$

where u(x) and v(x) are two scalar functions of the independent variable x over a domain and n_x is the component in the x direction of the outwardly directed normal to the surface of the domain. It is then necessary to express the heat flux by Fourier's equation, and to make the approximation implied by the interpolating function

$$t = \sum_j N_j t_j \quad j = 1,8 \tag{6}$$

from which the pertinent derivatives of t with respect to r and to z can be obtained. Performing the above operations and inserting the results into equation (4) gives an assembly of eight equations for each node i of the form

$$\sum \underline{KT}_{ij} t_j = \underline{RA}_i \quad j = 1,8 \tag{7}$$

where $\underline{KT}_{ij}$ is an involved area integral, the details of which are given in Wills [5], and $\underline{RA}_i$ is the sum of the two dimensional generation integral and the surface flux integral. When assembled, the terms form an overall matrix equation that can be written as $K\overline{T} = \underline{R}$ where K is the conductance stiffness matrix, $\overline{T}$ is the unknown vector formed by the temperatures at the individual nodes, and $\underline{R}$ is the matrix representing net energy leaving the element due to known generation and external heat flux boundary conditions. The heat flux on the internal element faces is equal and opposite in sign and therefore cancels during the assembly process. The enormous amount of detail regarding assembly and coding of the matrix equations is thoroughly described by Wills [5].

4.2 The Displacement Element Equations

Most published formulations of the element equations for displacement and stress analysis utilize the minimum potential energy principle. The formulation herein is begun from an entirely different viewpoint; that of the Navier equilibrium equation for a homogeneous, isotropic, nonisothermal media:

$$(\lambda + 2\mu)\nabla(\nabla \cdot \overline{\delta}) - \mu\nabla x(\nabla x\overline{\delta}) + \overline{F} - \beta\nabla t = 0 \tag{8}$$

In this expression $\overline{\delta}$ is the displacement vector in cylindrical coordinates, $\overline{F}$ is the body force per unit volume, the quantity β is a material property related to thermal expansion coefficient, α, by $\beta = (3\lambda+2\mu)\alpha$ where λ and μ are Lame's constants. Applying the axisymmetric restriction to equation (8) and expanding the vector operator, ∇, yields for the r and z directions

$$(\lambda + 2\mu)[u_{rr} + (u/r)_r] + \lambda w_{zr} + \mu(w_r + u_z)_z + F_r - \beta\partial t/\partial r = 0 \tag{9}$$

and

$$(\lambda + 2\mu)[w_{zz} + (u_r + (u/r))_z] + \lambda(u_z + w_r)_r$$

$$+ (\mu/r)[u_z + w_r] + F_z - \beta\partial/\partial z = 0 \tag{10}$$

where the subscript notation for partial differentiation has been used to conserve space, i.e. $w_{zz} = \partial(\partial w/\partial z)/\partial z$. The task now requires application of Galerkin's method as indicated by equation (3) to equations (9) and (10) using again as a domain the eight-node quadrilateral toroidal element and as approximating functions N_i, only we now write $N_i = N^i$ (no tensor flavor intended) to accommodate our subscript notation for partial differentiation; i.e., $N_r^i = \partial N_i/\partial r$. Using integration by parts permits writing the resulting application of Galerkin's method to equation (9) as

$$\int_A \{(\lambda+2\mu)[rN_r^i u_r + uN^i/r] + \lambda(N^i u_r + uN_r^i) + \mu rN_z^i(u_z + w_r) +$$

$$\lambda w_z(N^i + N_r^i) - rN^i F_r - \beta(N^i + rN_r^i)t\}drdz =$$

$$\int_S rN^i\{[(\lambda+2\mu)u_r + \lambda(u/r + w_r) - \beta t]n_r + [\mu(w_r + u_z)]n_z\}ds \tag{11}$$

and when applied to equation (10), which relates to displacement in the axial direction, we obtain

$$\int_A \{(\lambda+2\mu)rN_z^i w_z + \lambda r[N_r^i u_r + uN_z^i/r] + \mu rN_r^i(u_z + w_r) -$$

$$r(\beta N_z^i t + N^i F_z)\}drdz =$$

$$\int_S rN^i\{[(\lambda+2\mu)w_z + \lambda(u/r + u_r) - \beta t]n_z + [\mu(u_z + w_r)]n_r ds \tag{12}$$

The reader will note the distinction between subscripts r and z denoting differentiation and the olde english subscripts $\mathfrak{r}$ and $\mathfrak{z}$ denoting radial and axial direction. Let $\hat{R}$ and $\hat{Z}$ be the components of the surface tractions of the domain such that

$$\hat{R} = \sigma_r n_r + \tau_{r\mathfrak{z}} n_{\mathfrak{z}} \tag{13}$$

$$\hat{Z} = \tau_{r\mathfrak{z}} n_r + \sigma_{\mathfrak{z}} n_{\mathfrak{z}} \tag{14}$$

where Timoshenko and Goodier [8] or Boley and Wiener [9] have shown that

$$\sigma_r = (\lambda + 2\mu)u_r + \lambda(w_z + u/r) \tag{15}$$

$$\sigma_{\bar{z}} = (\lambda + 2\mu)w_z + (u_r + u/r) \tag{16}$$

$$\tau_{r\bar{z}} = \mu(u_z + w_r) \tag{17}$$

Therefore the wavy bracketed terms in the surface integrals of equations (11) and (12) can be expressed in terms of the surface traction components so that substitution yields

$$\int_A \{- - - - - -\}drdz = \int_s rN_i\hat{R}ds \tag{18}$$

$$\int_A \{+ + + + + +\}drdz = \int_s rN_i\hat{Z}ds \tag{19}$$

The terms in the area integrals of equations (18) and (19) have for convenience not been repeated. The final steps in the displacement element equation formulation requires the approximation of r, u, w, t, u_r, u_z, w_r, and w_z by the interpolation functions N_i and substitution into equations (18) and (19) for each node in the domain. The assembly of terms from the area integrals can be separated into the stiffness matrix $\underline{K}$ and the vectors of unknown displacements $\bar{\delta}$; the terms from the surface integral comprise the matrix $\bar{R}$ of body; thermal expansion, and impressed forces such that $\underline{K}\bar{\delta} = \bar{R}$.

The above abbreviated account completes the formulation of the element equations in global coordinates r and z. Evaluation of the integrals appearing in the terms for $\bar{R}$ and $\underline{K}$ in local coordinate systems was accomplished by use of three point Gaussian quadriture as described by Carnahan, Luther, and Wilkes [10].

After evaluating the integrals of the various terms numerically one must apply external boundary conditions to eventually arrive at the final matrix equations; we now turn our attention to a description of those boundary conditions.

5 BOUNDARY CONDITIONS

The surfaces at which boundary conditions must be applied are shown in Fig. 3. Surfaces 1, 3 and 5 are symmetry planes and are adiabatic with zero stress. Surface 4 is the outer surface of the fuel rod that is in contact with liquid sodium at 1000F with a unit surface conductance of 30000 $B/hrft^2F$ and zero radial and axial stress. Interstitial surfaces 6, 7, and 8 have thermal gap conductance and no frictional coupling. No frictional coupling was considered because at the power levels investigated thermal expansion was not enough to close the gap between the fuel and washer or the gap between the fuel and cladding. For surfaces 6 and 8 the fuel-clad and the washer-clad gap conductances were set at 1500 $B/hrft^2F$; and for surface 7 the fuel-washer conductance was taken to be 3000 $B/hrft^2F$.

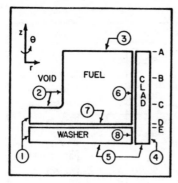

Fig. 3. Region boundaries

The selection of these values is extensively discussed in References [3] and [4], and represents values indicative of actual operating conditions.

Surface 2 contributes the major nonlinearity aspect of
this work; physical property variation with temperature also
contributes to the nonlinearity of the problem. The radiation
boundary condition on the centervoid (surface 2) involves the
description of the interchange factors between the nodal surfaces on the centervoid, a very complicated problem in itself,
and the interaction between irradiation and radiosity at the
nodal surface. The procedure for handling the radiation nonlinearity is described at length in the next section.

7 THE RADIATION BOUNDARY CONDTIION

All surfaces emit energy at a rate proportional to the
fourth power of their absolute temperature. If the interior
surface of an enclosure is not isothermal, there is a net energy exchange from the higher to the lower temperature portions
of the surface. Assume that the surface subject to radiation
heat transfer can be divided into n subsurfaces, each defined
by a node in the finite element discretation process, as shown
in Figure 4, and that each of these subsurfaces is diffuse and
opaque. Further assume that the emmissivity, ε, the reflectivity, ρ, and the temperature t, can be considered as invariant
over the extent of the subsurface. The total heat flux per
unit area into subsurface j can then be expressed as:

$$q_j^* = G - J \tag{20}$$

where G is the total irradiation of subsurface j by all other
subsurfaces on the enclosure and J is the total radiosity from
subsurface j. Since temperature and surface properties are
assumed constant on each of the n subsurfaces,

$$q_j^* = \sum_{i=1}^{m} J_i F_{ij} A_i - J_i A_j \tag{21}$$

80

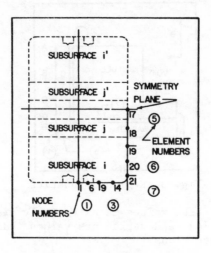

Fig. 4. Enclosure
subsurface definitions

where F_{ij} is the configuration factor from subsurface i to subsurface j and A_i and A_j are the areas of subsurfaces i and j,
respectively. Casting (21) into matrix form yields

$$\underline{H} = \underline{L} \, \underline{J}$$

where
$$\underline{H} = q_j \quad ; \quad \underline{J} = J_j$$

and
$$\underline{L} = \begin{bmatrix} F_{ji}A_j & , & i \neq j \\ A_i(F_{ii}-1), & i = j \end{bmatrix} \tag{22}$$

From the definition of radiosity,

$$J_j = \varepsilon_j \sigma t_j^4 + (\rho_j/A_j) \sum_{\substack{i=1 \\ i \neq j}}^{n} J_i F_{ij} A_i \tag{23}$$

where σ is the Stefan-Boltzmann constant, 5.669×10^{-8} W/m^{2}K^4, ε
and ρ are the surface emissivity and reflectivity, and t_j^4 is
the absolute temperature of subsurface j raised to the fourth
power. We may write in matrix form

$$\underline{D} \, \underline{T}^4 = \underline{M} \, \underline{J} \tag{24}$$

where

$$\underline{D} = \begin{bmatrix} \varepsilon_i & , & i = j \\ 0 & , & i \neq j \end{bmatrix} \tag{24a}$$

$$\underline{T}^4 = t_i^4 \tag{24b}$$

$$\underline{M} = \begin{bmatrix} -F_{ji} \; _iA_j/A_i & , & i \neq j \\ 1 - \rho_i F_{ii} & , & i = j \end{bmatrix} \tag{24c}$$

Solving (24) for J and substituting into (22) yields

$$\underline{H} = \underline{L} \; \underline{M}^{-1} \; \underline{D} \; \underline{T}^4 \tag{25}$$

the matrix equations (24) and (25) can be written in the short-hand form,

$$\underline{Q} = \underline{B} \; \underline{T}^4 \tag{26}$$

where

$$\underline{B} = \underline{A} \; \underline{L} \; \underline{M}^{-1} \; \underline{D} \tag{27}$$

and

$$\underline{A} = \begin{bmatrix} -1/A_i & , & i = j \\ 0 & , & j \neq j \end{bmatrix} \tag{28}$$

and, in particular, $Q = q_j^*$, the heat flux per unit are out of subsurface j, the value for substitution into the boundary condition term in the finite element matrix equation (4).

The square matrix $\underline{B}$ depends upon material surface properties, the geometry under consideration, and the choice of domain discretation. Unfortunately, the matrix $\underline{T}^4$ is a vector of the absolute temperatures raised to the fourth power of all nodes on the enclosure surface, in all likelihood, unknown. The application of the radiation boundary condition, thus, will require an iterative process. It should be noted that should the enclosure be black ($\varepsilon = 0$, $\rho = 1$), the matrix $\underline{M}$, as defined by (24), becomes the identity matrix and, as such, need not be inverted. Direct multiplication gives the matrix immediately as

$$\underline{B} = \begin{bmatrix} -\sigma F_{ji} \; _iA_j/A_i & , & i \neq j \\ \sigma(1 - F_{ii}) & , & i = j. \end{bmatrix} \tag{29}$$

When such an assumption can be made it greatly reduces the computational effort involved.

To complete the algorithm requires detailed knowledge of the interchange factors F_{ij} and F_{ii} contained in $\underline{B}$; the next section summarizes the calculation method described in complete detail by Wills [5].

8 GENERALIZED AXISYMMETRIC CONFIGURATION FACTORS

The development above for the implementation of the radiation boundary condition is applicable to any shape internal enclosure. For arbitrary shapes, however, the calculation of the configuration factors can be considerably simplifed.

Assume there are two subsurfaces on such an enclosure, A and B, for which the configuration factor F_{AB} is desired, and that these subsurfaces were formed by the rotation of two non-coincidental arcs A and B in r, θ, z cylindrical coordinates about the common z axis as shown in Figure 5.

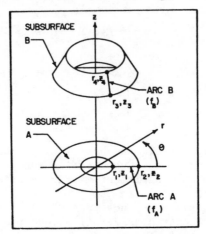

Fig. 5. Geometrical arrangement
of subsurfaces A and B in
cylindrical coordinate space r, θ, z.

Let arc A be given by $z = f_A(r)$ and arc B by $z = f_B(r)$, where f_A and f_B are arbitrary funcations, independent of the circumferential coordinate, θ. Further, let arc A have endpoints (r_1, z_1) and (r_2, z_2) and arc B have endpoints (r_3, z_3) and (r_4, z_4). Four parallel directly opposed (possibly in part coincident) circular disks are then implicitly defined at $z = z_1$, $z = z_2$, $z = z_3$, and $z = z_4$ with radii of r_1, r_2, r_3 and r_4, respectively. Let these four disk surfaces by denoted 1,2,3, and 4.

Using flux algebra as outlined by Wolf [11], the geometric flux from subsurface A to subsurface B is defined as

$$G_{AB} = A_A F_{AB} \qquad (30)$$

where A_A is the area of subsurface A viewable from subsurface B and F_{AB} is the configuration factor from subsurface A to subsurface B. We may then write

$$G_{AB} = G_{A3} - G_{A4} \qquad (31)$$

and with the use of ordinary reciprocity we may express G_{A3} and G_{A3} and G_{A4} as

$$G_{A3} = G_{3A} = G_{32} - G_{31} = G_{23} - G_{13} \qquad (32)$$

$$G_{A4} = G_{4A} = G_{42} - G_{41} = G_{24} - G_{14} \qquad (33)$$

Substituting (32) and (33) into (31) and utilizing the definition given in (30) we obtain

$$F_{AB} = (A_1/A_A)(F_{14} - F_{13}) - (A_2/A_A)(F_{24} - F_{23}) \qquad (34)$$

From the Theorem of Pappus, we can solve for the area,

and with
$$A_A = 2\pi \int_{r_1}^{r_2} r [1 + df_A/dr^2]^{.5} \, dr \qquad (35)$$

$$A_1 = \pi \, r_1^2 \text{ and } A_2 = \pi \, r_2^2,$$

we finally arrive at

$$F_{AB} = [r_1^2(F_{14} - F_{13}) - r_2^2(F_{24} - F_{23})]/$$

$$(2\int_{r_1}^{r_2} r[1 + (df_A/dr)^2]^{.5} dr) \qquad (36)$$

In order to evaluate F_{AB}, the configuration factor between two possibly complex totally visible surfaces, we need only determine the area of subsurface A and the configuration factors between the parallel, concentric, disk surfaces 1 and 4, 1 and 3, 2 and 4, and 2 and 3. These configuration factors have previously been evaluated and are given by Hamilton and Morgan [12] as

$$F_{ij} = [Y - (Y^2 - 4E^2D^2)^{.5}]/2 \qquad (37)$$

where
$$D = (z_j - z_i)/r_i \qquad (38a)$$

$$E = r_j /(z_j - z_i) \qquad (38b)$$

84

$$Y = 1 + (1 + E^2) \ D^2 \qquad (38c)$$

Substituting equations (37) and (38) into equation (36) yields

$$F_{AB} = [z_A z_B + \sum_{i=1}^{2} \sum_{j=3}^{4} (-1)^{i+j} X_{ij}]/$$

$$(2\int_{r_1}^{r_2} r[1 + (df_A/dr)^2]^{.5} \ dr) \qquad (39)$$

where

$$z_A = z_2 - z_1$$

$$z_B = z_4 - z_3$$

and

$$X_{ij} = .5 \ \{[r_i^2 + r_j^2 + (z_i - z_j)^2]^2 - 4 \ r_i^2 \ r_j^2\}^{.5}$$

When arc A is a line segment (a fairly common geometric arrangement) such that df_A/dz is a constant, equation (39) can be specialized to obtain a useful working equation as follows

$$F_{AB} = [z_A z_B + \sum_{i=1}^{2} \sum_{j=3}^{4} (-1)^{i+j} X_{ij}]/(r_2 + r_1)(r_A^2 + z_A^2)^{.5} \qquad (40)$$

where $r_A = r_2 - r_1$ and z_A as above. The formulation described by equation (40) permits calculation of the interchange factor for a wide variety of axisymmetrical geometries such as disk to washer, disk to cylindrical ring, disk to cone segment, and so forth.

When subsurface A is reentrant and capable of intercepting its own emitted radiation (such as a cylindrical ring or cone segment), equation (39) will not directly apply. In such a case the reflexive configuration factor, F_{AA}, is required. It can be shown, again using flux algebra, that by equating r_1 and r_3, z_1 and z_3, r_2 and r_4, and z_2 and z_4, using an appropriate form of equation (39) to calculate a factor F'_{AA}; then the reflexive configuration factor is given by

$$F_{AA} = 1 - F'_{AA} \qquad (41)$$

8.1 Effect of Symmetry Planes

Quite often when dealing with axisymmetric enclosures, symmetry also exists about an r-θ plane. In such cases, the associated conduction problem in the region surrounding the enclosure can be handled by assigning an adiabatic boundary condition at the plane of symmetry and analyzing only one half of the continuum. A similar procedure may be employed to considerable advantage to simplify the radiation problem within symmetric axisymmetric enclosures.

Consider the enclosure shown in Figure 4 where subsurface i' is the symmetrical counterpart of subsurface i. While these two subsurfaces are both geometrically and thermally equivalent, the energy transfer from i to another subsurface j will be quite different than that transferred to j from i', due to the difference in the configuration factors F_{ij} and $F_{i'j}$. The situation may be handled in the confines of the symmetrical section by letting the two subsurfaces i and i' form a composite subsurface I, and, instead of computing the energy transfer from i to j, compute the energy transfer from I to j. The computation is easily accomplished by replacing the configuration factors, F_{ij}, in the matrices $\underline{L}$ and $\underline{M}$ (equations (22) and (24) by the psuedo factor

$$F^*_{Ij} = F_{ij} + F_{i'j} \tag{42}$$

whether or not the subsurface i is reentrant, the composite surface I is and, thus, using equation (41)

$$F^*_{Ii} = 1 - F'_{ii} + F_{i'i} \tag{43}$$

where F'_{ii} is obtained using equation (39) with coincident arcs. Having calculated the set of pseudo factors, those subsurfaces forming the redundant portion of the symmetric enclosure need no longer be considered, and one can proceed as though the symmetry plane, as it passes through the enclosure, consists of a perfectly absorbing, non-emitting surface.

9 SOLUTION OF THE ENCLOSURE RADIATION PROBLEM

With the aid of the above development, one is able to proceed with the automation of the solution to nonlinear radiation problems in axisymmetric enclosures. The first step, as in all numerical analyses, is to discretize the domain. Once nodes have been established on the surface subject to radiation heat transfer, let each node define a subsurface, bounded on either side at a point midway between the defining node and the adjacent node. If no node exists on one side or the other, the defining node itself bounds the subsurface. It was found convenient, although introducing a slight error, to let these subsurfaces be formed by line segments connecting the nodes such that with the use of the discretization data and equations (40), (42) and (43) a matrix of configuration factors could be immediately calculated. In most instances (there are exceptions such as when large deformations occur) this matrix is invariant for a given discretation.

Next, proceed in the solution as though the enclosure surface were adiabatic, i.e., set up the matrix equation $\overline{KT} = \underline{R}$ with q = 0 for those nodes on the enclosure surface. The solution to this linear set of equations provides an excellent choice of starting temperatures for the iteration process. If nonlinear material properties are also included in the problem, some nominal value of temperature may be chosen for this first iteration.

Then, using the temperatures so calculated, evaluate the matrix $\underline{Q}$ using equation (26) along with the matrix of configuration factors. Recalculate the matrix equation $\overline{KT} = \underline{R}$ with subsequent values of $\underline{Q}$ inserted for q at the appropriate nodes subject to radiation and solve simultaneously with $\overline{K\delta} = \underline{R}$. Since the thermal problem depends on the stress evaluation only through the possibility of the specialized boundary condition of gap conduction the most straight forward approach was by first solving the thermal equation, then the stress equation and iterating the procedure until convergence was achieved. The iteration procedure was selected because of significant advantages in computer storage space and execution time.

The program AXIFE (AXIsymmetric Finite Element code) was written by Wills [5] to implement solution to the matrix equations formulated above. Input data to the program is used to define: the domain extent, imposed boundary condition, the desired domain discretization, and the material properties. Since almost any axisymmetric domain may be modeled with the modification of the input data, the usefulness of the code AXIFE is extended far beyond the context of the subject investigation.

The algorithms in the code were verified by performing a thermal and stress analysis on a hollow sphere and comparing the results with the closed form solutions given by References [7] and [8]; the comparison showed excellent agreement with differences between exact and code results in the range of one to two percent.

The code was set up to handle 190 nodes for a stress analysis and 380 nodes for thermal analysis, and altogether requires 864K bytes of computer storage. For 104 nodes each thermal analysis iteration for the composite fuel module required 1.2 minutes, and 3.2 minutes for a stress iteration. The 6 kw/ft runs described in the results section took about 25 to 30 minutes on the average on an Amdahl 470 V/5 system. Total execution time strongly depends on the number of nodes,

the type of boundary conditions, the starting values used in
the iteration, and the convergence criteria set.

In addition to comparing code AXIFE results for a sphere
with the exact solution, several checks were made to make sure
thermal and stress results were realistic. On all results, a
First Law check was made to be sure that the energy generated
in the fuel volume was equal to the heat flux leaving the surface of the cladding, and that the summation of the net radiant
fluxes on all subsurfaces forming the centervoid was equal to
zero. For the stress results an equilibrium check was made to
be sure that element average stress values tend toward those
values of surface tractions which prevent rigid body notion.

The objectives of this work were to investigate the effect
of thermal radiation in the centervoid and to evaluate the
attendant thermal stresses; we now describe some typical results.

10 THERMAL RESULTS

Thermal results were computed for the range of rod power
from 3 kw/ft to 24 kw/ft; however, at values greater than
6 kw/ft severe stress cracking was indicated so that no extensive thermal investigations were carried out at the higher
power levels. Since previous work had been done by Krudener
[4] and Almond [6] using finite difference techniques at a
level of 18 kw/ft with an adiabatic centervoid ($\varepsilon=0$) it was of
interest to make a calculation with the finite element method
described herein for 18 kw/ft and $\varepsilon=0$ to compare temperatures
at the hot spot and other locations. For the geometry discussed in this work Krudener calculated a maximum fuel temperature
of 4566R at the symmetry plane on the centervoid surface; the
present code AXIFE predicts a temperature of 4580R at the same
location; a value that is some 14 degrees higher and represents
very satisfactory agreement. In general the agreement between
the two investigations is good with a maximum difference of
only five percent at the location where the fuel, washer, and
cladding are contiguous.

The variation of temperature with radial position in the
fuel module is illustrated in Figure 6 which shows that as
emissivity increases from zero (adiabatic centervoid) to a
maximum possible value of 1.0 (black surface) the maximum fuel
temperature decreases and also moves outward radially in the
fuel in the direction toward the cladding. At an emissivity of
0.9 the maximum fuel temperature is about one third of the fuel
radial thickness in from the centervoid surface and has been
reduced to 4407R as compared to the adiabatic maximum of 4580R.
The value of $\varepsilon=0.9$ has been selected as being a realistic
approximation of surface radiation characteristics. According-
ly, Figure 7 delineates the variation of the temperature in the
fuel module with radial position at 18 kw/ft and $\varepsilon=0.9$; the

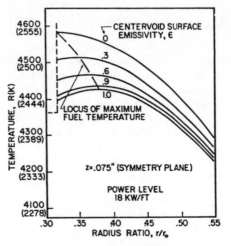

Fig. 6. Effect of void surface emissivity on
location and magnitude of maximum fuel temperature

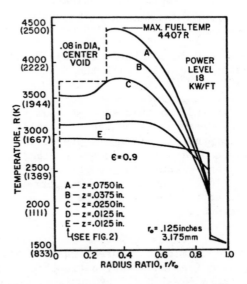

Fig. 7. Variation of temperature with radial
position in the fuel module with centeroid radiation
present, ε=0.9 and 18 kw/ft rod power

curves labeled A through E are at the axial locations shown in
the figure. Curves A, B, and C all show that the maximum temperatures in the fuel have all shifted into the fuel in the
direction of the cladding and curve C, which is the elevation
(axial position) of the bottom horizontal centervoid surface,
shows a minimum at about a position ratio of about 0.2 due to

the small 0.010 inch diameter hole on the centerline. Consequently the energy flow pattern near the centerline is somewhat
more complicated than in the adiabatic centervoid case without
the small hole present. The temperature profile represented by
curve E (in Figure 7) is that at the washer side of the washer-
fuel gap and is only slightly higher than the adiabatic centervoid case; therefore, more of the energy generated in the fuel
flows ultimately through the washer into the cladding. The
increase in energy flux through the washer results in a maximum
washer temperature of 2962R for the .9 emissivity case as opposed to 2922R for the adiabatic case.

Figure 8 demonstrates the effect of rod power on the magnitude and location of the maximum fuel temperature; only results for the central portion of the fuel pellet are shown for

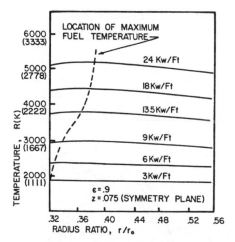

Fig. 8. Effect of power level on fuel
temperature near the maximum on the
symmetry plane for $\varepsilon=0.9$

centervoid surface $\varepsilon=0.9$. The figure shows that the proposed
geometry is thermally capable of handling 24 kw/ft with all
portions of the fuel remaining under the estimated 5100R melting temperature [13] and is relatively flat near the maximum.
At the outer edges of the fuel the temperature gradients are so
steep that at rodpowers over 6 kw/ft the stresses induced far
exceed the tensile strength of the fuel; stress aspects are
discussed at length in the next section. Figures 9 and 10 show
the influence of emissivity on the horizontal and vertical surface temperatures respectively of the centervoid at 18 kw/ft.
The figures show that the net effect as emissivity is increased
from the adiabatic case to a more realistic value (such as .9)
is that the void surface becomes more nearly isothermal, albeit
less so on the vertical surface. Radiant heat transfer from
the higher temperature regions near the symmetry plane to the
lower temperature horizontal void surface provides an alternate

90

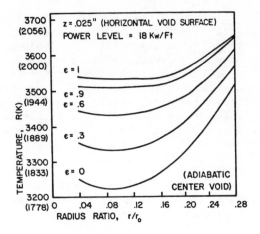

Fig. 9. Influence of emissivity on the
horizontal centervoid surface temperature at 18 kw/ft

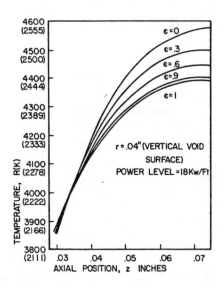

Fig. 10. Influence of emissivity on the
vertical centervoid surface temperature at 18 kw/ft

transport path for energy generated in the higher temperature
regions of the fuel.

As was mentioned above, rod powers above six kw/ft cause
such severe tangential thermal stresses in the outer edges of
the fuel that normal operation would cause severe cracking and
possible spalling of the fuel. In themselves, radial cracks

do not inhibit the flow of thermal energy to the cladding; they are undesirable from the standpoint of possible spalling and fuel redistribution. All current annular fuel designs show tangential stress cracks (radial) under normal operating conditions. See for example, the artists rendition on the cover of reference [141 as a graphic illustration of the happenings. We now turn our attention to the description of the stress results.

11 STRESS RESULTS

At moderately high power levels elastic axisymmetric stress analysis predicts tangential stresses on the outer surface of the fuel which far exceed the tensile strength of the ceramic. The magnitude of these tangential stresses along with those obtained in the analysis of the conventional annular fuel are shown in Figure 11. It is evident, as mentioned above, that in such a case, when tangential stresses exceed the tensile strength of the material, radial cracks, extending from

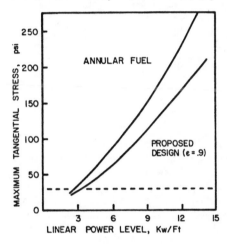

Fig. 11. Comparison of maximum tangential
stresses predicted for the proposed design
and for annular fuel under elastic conditions

the outer surface of the fuel toward the centerline are initiated. Unfortunately, such fractures, while in theory not affecting the temperature distribution, introduce a circumferential dependence on the stress field which cannot be accommodated in an axisymmetric analysis and, thus, invalidates the results obtained by using the geometrical constraints implied by such an assumption. Guha, Head, and Matthews [15], predicting such fracture formation in annular fuel, handled the problem by utilizing a two-dimensional plane stress model in the r - θ plane. In the subject analysis, however, due to axial nonuniformity, a three-dimensional code would be needed for analysis, requiring, for accuracy a presently unobtainable amount of computer storage space.

Thermal stress analyses at fairly low power levels, showed a significant decrease in the maximum tangential stresses in the fuel. At 6 kw/ft it is felt that cracking would be confined to the outer surface of the fuel, thereby permitting reasonably accurate stress predictions from code AXIFE. Figure 12 presents such tangential stress results at a power level of 6 kw/ft and $\varepsilon=.9$ in the centervoid and shows that the tensile strength is exceeded only at radius ratios greater than about 0.7; the centervoid surface is shown to be in compression.

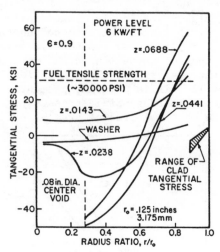

Fig. 12. Variation of tangential stresses radial position in the fuel module at 6 kw/ft and $\varepsilon=.9$

Figures 13 and 14 show the variations of radial and axial

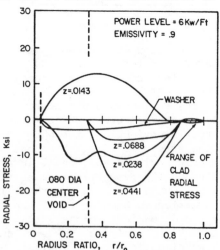

Fig. 13. Variation radial stresses with radial position in the fuel module at 6 kw/ft and $\varepsilon=0.9$

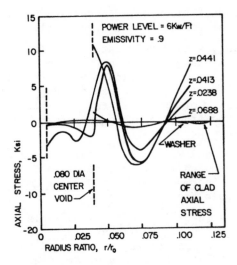

Fig. 14. Variation of axial stresses with radial
position in the fuel module at 6 kw/ft and ε=0.9

stresses (respectively) in the fuel module at 6 kw/ft and a
surface emissivity of 0.9 in the centervoid. The relationships
shown are complex, but none of the stresses exceed 18 ksi and
most are less than 10 ksi. Reflection on the stress results
clearly indicates that the fuel pellet must be stress relieved
at the outer edge to alleviate the very large tangential
stresses encountered at rod power levels where the thermal capabilities of the proposed fuel module design could be more
fully utilized. Such stress relief could be obtained by casting very thin radial slots into the fuel to an optimum depth
and in an optimum geometrical arrangement in the θ direction.
Unfortunately, such an optimization requires a three dimensional finite element code and was beyond the capabilities of
AXIFE.

12 ACKNOWLEDGEMENTS

The code AXIFE was constructed and the analysis was performed by the first author as the major portion of his dissertation, the paper was written by the second author who served
as dissertation advisor, and was very competently typed by Mrs.
Betsy Cole. Financial support was provided by the University
of Arkansas Computer Center, the Mechanical Engineering Department, and Arkansas Power and Light Company of Little Rock,
Arkansas.

13 REFERENCES

[1] MASON, W.E.; "Finite Element Analysis of Coupled Heat
    Conduction and Enclosure Radiation", Proceedings of the
    International Conference on Numerical Methods, Swansea,
    U.K., July 1979.

94

[2] KARAM, R.D., and EBY, R.J., "Linearized Solution of Conducting--Radiating Fins", AIAA Journal, Vol. 16, No. 5, 1978.

[3] KRUDENER, R.M., A Matrix Iterative Development of Transient and Steady Temperature Distributions in Nuclear Fuel Elements Incorporating an Auxiliary Heat-Transport Path Concept, Ph.D. Dissertation, Mechanical Engineering Department, University of Arkansas, Fayetteville, Arkansas (1973), (also available from Library of Kernforschungszentrum, D75 Karlsruhe, West Germany).

[4] KRUDENER, R.M. and WOLF, H., "Prediction of Steady and Transient Thermodynamic Temperatures in Composite Nuclear Fuel Rods" in Thermodynamics of Nuclear Materials, Vol. I, Internat. Atomic Energy Agency, Vienna, Austria 1975, p.45.

[5] WILLS, L.D., A Thermal and Stress Analysis of A Composite Nuclear Fuel Rod Using Nonlinear Finite Element Techniques, Ph.D. Thesis, Mechanical Engineering and Engineering Science Department, University of Arkansas, 1982.

[6] WOLF, H., WILLS, L.D., KRUDENER, R.M., and ALMOND , M.D., "Thermal and Stress Analysis of Composite Nuclear Fuel Rods by Numerical Methods", Chapter 8 in Numerical Methods in Heat Transfer, edited by Lewis, Morgan, and Zienkiewicz, John Wiley and Sons, 1981.

[7] ZIENKIEWICZ, O.C., The Finite Element Methods, 3rd ed., McGraw-Hill, London, UK, 1977.

[8] TIMOSHENKO, S.P., and GOODIER, J.N., Theory of Elasticity, 3rd ed., McGraw-Hill, New York, New York, 1970.

[9] BOLEY, B.A., and WIENER, J.A., Theory of Thermal Stresses, John Wiley and Sons, New York, 1960.

[10] CARAHAN, B., LUTHER, H.A., and WILKES, J.O., Applied Numerical Methods, John Wiley and Sons, New York, New York, 1969.

[11] WOLF, H., Heat Transfer, Harper & Row Publishers, New York, N.Y. 1982, p.359.

[12] HAMILTON, D.C., and MORGAN, W.R., "Radiant Interchange Configuration factors", NACA Report No. TN 2836, Dec. 1952.

[13] AITKEN, E.A., EVANS, S.K., A Thermodynamics Data Program Involving Plutonia and Urania at High Temperatures, Gen. Elec. Co., GEAP-5634.

[14] DALLEDONNE, M., KUMMERER, K., and SCHROETER, K., editors, Fast Reactor Fuel and Fuel Elements, Proceedings of an International meeting Kernforachungszoutrun, Karlsruhe, FRG, 1970, published by Gesellschaft fuer Kernforschung mbH., Karlsruhe, FRG.

[15] GUHA, R.M., HEAD, J.L., and MATHEWS, J.R., "Analysis of Crack Patterns in Fast Reactor Fuel Pellet", Nuclear Energy, Vol. 18, 1979.

CHAPTER 4

A NUMERICAL METHOD FOR COMPUTING MASS AND ENERGY TRANSPORT
THROUGH PARTIALLY SATURATED POROUS MEDIA*

Roger R. Eaton
Fluid Mechanics and Heat Transfer Division I
Sandia National Laboratories
Albuquerque, New Mexico   87185

1. SUMMARY

This paper discusses the development of the finite element
computer code SAGUARO which calculates the two-dimensional flow
of mass and energy through porous media. The media may be saturated or partially saturated. SAGUARO solves the parabolic
time-dependent mass transport equation which accounts for the
presence of partially saturated zones through the use of highly
non-linear material characteristic curves. The energy equation
accounts for the possibility of partially-saturated regions by
adjusting the thermal capacitances and thermal conductivities
according to the volume fraction of water present in the local
pores. The code capabilities are demonstrated through the
presentation of a sample problem involving the one-dimensional
calculation of simultaneous energy transfer and water infiltration into partially saturated hard rock.

2. NOMENCLATURE

The nomenclature used in this paper are given below.

		Typical Units
c_p	- specific heat	J/kg•K
C	- derivative of moisture content $(\partial\theta/\partial\phi)$	m^2/N

---

*This work performed at Sandia National Laboratories supported
by the U.S. Department of Energy under contract number DE-
AC04-76DP00789. The work described in this paper is intended
to contribute towards a general understanding of the hydrology
in the region of porposed high-level nuclear waste underground
repositories. The funding for this work was provided by the
Nevada Nuclear Waste Storage Investigations Project managed by
the Nevada Operations Office of the U.S. Department of Energy.

The author acknowledges D. K. Gartling and D. E. Larson for
their assistance on this problem.

D_{ij}	–	thermal diffusion tensor (Soret)	$m^2/s \cdot K$
E_{ij}	–	thermal dispersion tensor	$J/s \cdot m \cdot K$
g	–	gravitational constant	m/s^2
k_{ij}	–	intrinsic permeability tensor	m^2
p	–	liquid (pore) pressure	N/m^2
Q	–	volumetric heat generation rate	$J/m^3 \cdot s$
t	–	time	s
T	–	temperature	K
v	–	Darcy velocity (mean specific flux, $m^3/s \cdot m^2$)	m/s
x	–	horizontal dimension	m
z	–	vertical dimension	m
α	–	acquifer compressibility	m^2/N
β	–	coefficient of volumetric thermal expansion	$kg/m^3 {}^\circ C$
Γ	–	water compressibility	m^2/N
Δ	–	increment	----
θ	–	moisture content	----
λ_{ij}	–	thermal conductivity tensor	$J/s \cdot m \cdot K$
μ	–	dynamic viscosity	$kg/m \cdot s$
ρ	–	liquid density	kg/m^3
Φ	–	hydraulic head = $\rho g(\psi + z)$	N/m^2
ϕ	–	porosity	----
ψ	–	pressure head, $p/(\rho g)$	m

2.1 Subscripts

eff – effective property
f   – fluid
i,j – coordinate direction
o,r – reference condition
s   – solid property

3. INTRODUCTION

The Nevada Nuclear Waste Storage Investigations Project, managed by the Nevada Operations Office of the U.S. Dept. of Energy, is examining the feasibility of siting a repository for high-level nuclear wastes at Yucca Mountain on and adjacent to the Nevada Test Site. The hydrological aspect of this problem has prompted the development of a computational code, SAGUARO [1]. SAGUARO is a general-purpose finite element code developed to solve time-dependent problems of single phase water and energy transport through porous media which may be partially or fully saturated. The two transport equations (mass and energy), which model the flow, incorporate Darcy's law, the Boussinesq approximation, the Soret effect, conduction and convection. The resulting nonlinear parabolic equations are solved in finite-element form using an algorithm related to the standard Crank-Nicolson method. The matrix solution procedure used in SAGUARO is a form of Gauss elimination. Code results provide time and space distributions of hydraulic head, temperatures, velocities and moisture contents.

4. MATHEMATICAL FORMULATION

The present analysis will be restricted to the flow of an incompressible Newtonian fluid through a homogeneous, rigid porous matrix. The mass transport equation used in SAGUARO is derived using the equation for the continuity of mass

$$\frac{\partial(\rho_f v_i)}{\partial x_i} + \frac{\partial(\rho_f \theta)}{\partial t} = 0 \tag{1}$$

and the Darcy equation to define the velocity for laminar fluid flow,

$$v_i = \frac{-k_{ij}}{\mu}\left(\frac{\partial p}{\partial x_j} + \rho_f g \frac{\partial z}{\partial x_j}\right) - D_{ij}\frac{\partial T}{\partial x_j} \quad . \tag{2}$$

Substituting Equation (2) into (1) gives:

$$\frac{\partial}{\partial x_i}\left(\rho_f\left[-\frac{k_{ij}}{\mu}\left(\frac{\partial p}{\partial x_j} + \rho_f g \frac{\partial z}{\partial x_j}\right) - D_{ij}\frac{\partial T}{\partial x_j}\right]\right) = \frac{-\partial(\theta\rho_f)}{\partial t} \quad . \tag{3}$$

The indices (i,j) range over the values $(1,2)$ and summation is implied by repeated indices. Since the air is assumed to be free to escape at the surface the air pressure is small in comparison to the pressure in the liquid. Thus the water pressure, p, is simply the negative of the capillary pressure P_c:

$$p = -P_c(\theta) \quad . \tag{4}$$

Further neglecting hysteresis in the capillary pressure-moisture content relation, (4) can be inverted to give moisture content as a function of pore pressure or pressure head, which ever is desired. This observation, along with the Boussinesq approximation for thermal expansion of the fluid

$$\rho_f = \rho_0 (1 - \beta\Delta T) \quad , \tag{5}$$

and the assumption $\beta\Delta T \ll 1$, permits equation (3) to be rewritten as

$$\frac{\partial}{\partial x_i}\left(\frac{k_{ij}}{\mu}\frac{\partial\Phi}{\partial x_j}\right) - \frac{\partial}{\partial x_i}\left(\frac{k_{ij}}{\mu}\rho_0 g \beta\Delta T \frac{\partial z}{\partial x_i}\right) + \frac{\partial}{\partial x_i}\left(D_{ij}\frac{\partial T}{\partial x_j}\right)$$
$$= \frac{1}{\rho_0 g} C \frac{\partial\Phi}{\partial t} \quad . \tag{6}$$

In general, mass transport in partially saturated regions resulting from bouyancy effects is expected to be negligible. This effect is accounted for by setting $\beta \equiv 0$ when θ/ϕ is less 99%.

The right hand side of Equation (6) accounts for the time rate of change in mass being stored due to change in moisture content. Under high pressure, saturated conditions, additional phenomena such as aquifer compressibility and water compressibility may become significant. The storage term which accounts for these phenomena is

$$\text{Storage} = \frac{1}{\rho_0 g} \, S_s \, \frac{\partial \Phi}{\partial t} \quad ,$$

where

$$S_s = \rho_0 g (\alpha + \phi \Gamma)$$

For the general case the term which accounts for both partial saturation and high pressure compressibility effects becomes

$$\text{Total Storage} = \frac{1}{\rho_0 g} \, [\overline{C}(\psi)] \, \frac{\partial P}{\partial t}$$

where

$$\overline{C} = C + a S_s \quad .$$

By requiring "a" to be zero when the region is unsaturated and one when the region is saturated it is possible to analyze problems when compressibility is important in the saturated zone [2]. By replacing C by $\overline{C}$ in Equation (6), changes in moisture content, acquifier compressibility and water compressibility are accounted for. This is the mass transport equation solved for in SAGUARO.

It is of interest to note two special cases of equation (6). For isothermal flows, the second and third terms of equation (6) vanish, leaving

$$\frac{\partial}{\partial x_i} \left(\frac{k_{ij}}{\mu} \frac{\partial \Phi}{\partial x_j} \right) = \frac{1}{\rho_0 g} \, \overline{C} \, \frac{\partial \Phi}{\partial t} \quad . \tag{7}$$

This is the well-known Richards equation [3]. For saturated flows, the moisture content is constant ($\theta/\phi \equiv 1$) therefore, $\overline{C} \equiv 0$. Typically, mass transport resulting from temperature gradients, $\frac{\partial}{\partial x_i} \left(D_{ij} \frac{\partial T}{\partial x_j} \right)$, is much smaller than mass transport resulting from pressure gradients. For these conditions, the third and fourth terms in Equation (6) are neglected yielding a quasi-steady equation for flow through fully saturated media [4].

$$\frac{\partial}{\partial x_i} \left(\frac{k_{ij}}{\mu} \frac{\partial \Phi}{\partial x_j} \right) = \frac{\partial}{\partial x_i} \left(\frac{k_{ij}}{\mu} \rho_0 g \beta \Delta T \frac{\partial z}{\partial x_i} \right) \tag{8}$$

For partially saturated flow, the properties k_{ij} and C are strong functions of pore pressure. Thus Equation (6) is highly non-linear. Generally, the non-linear coefficients $k(\psi)$ and $\overline{C}(\psi)$ are functions similar to the typical wetting curves shown in Figure 1. It should be emphased that it is these curves which account for the influence of capillary action on pore water motion.

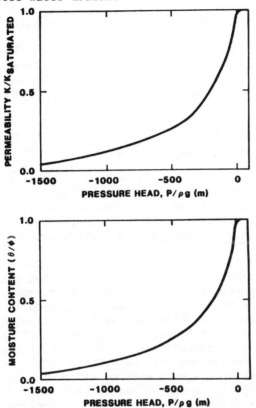

Figure 1. Characteristic curves relating permeability and moisture content to pressure head.

The general form of the energy equation solved by SAGUARO is

$$(\rho c_p)_{eff} \frac{\partial T}{\partial t} + \rho_{fo} \frac{\theta}{\phi} (c_p)_f v_i \frac{\partial T}{\partial x_i}$$

$$- \frac{\partial}{\partial x_i} \left[\left(\lambda_{eff_{ij}} - \phi E_{ij} \right) \frac{\partial T}{\partial x_j} \right] - Q = 0 \quad . \quad (9)$$

The equation represents an energy balance on a unit volume containing a porous matrix and liquid. It is assumed that the matrix and liquid are in thermal equilibrium and the air has negligible heat capacity. Both energy transport by

convection (2nd term) and conduction/dispersion (3rd term)
are included. The definitions of the material properties
are defined to allow for the possibility of partially
saturated pores by using mass weighted averages of heat
capacitances, densities and conductivities.

5. PROGRAM FEATURES AND ORGANIZATION

The usage of SAGUARO is restricted by the following
assumptions:
(a) The geometric description of the problem is limited to
     two spatial dimensions, either plane or axisymmetric.
(b) The matrix material may be considered orthotropic in
     terms of thermal conductivity, permeability and thermal
     diffusion.
(c) The matrix material(s) is assumed to be saturated or
     partially saturated with a single, one-phase fluid (no
     vapor transport). The fluid is assumed to be and
     Newtonian.
(d) The fluid flow is assumed to be laminar. Inertia
     effects in the fluid are assumed negligible.
(e) For nonisothermal flows, the fluid and material
     matrix are assumed to be in local thermal equilibrium.
(f) Presence of air in momentum/continuity equation is
     neglected.
Despite these restrictions, SAGUARO has proved to be a useful
tool in the solution of a wide range of transient porous flow
problems. When considering flows with heat transfer, regions
of solid-body heat conduction are easily included in the
analysis. Material properties such as fluid viscosity and
thermal conductivity may be arbitrary functions of temperature; volumetric heat sources may be functions of time and/or
temperature. Allowable boundary conditions on the hydrodynamic and thermal parts of a problem are quite general and
may include the specification of the fluid hydraulic head or
hydraulic head gradient (fluid discharge), as well as specified temperatures, heat fluxes or convective and radiative
boundaries. All boundary conditions may be functions of time.
Flow through fractured porous media is simulated by con-
structing the element mesh to include zoning for discrete
fractures. The saturated permeability in these zones is
determined by assuming the flow in the fracture to be
equivalent to water flow between parallel flat plates.

SAGUARO is a self-contained program with its own mesh
generator, data analysis, and plotting packages. The elements
included in SAGUARO consist of isoparametric and subparametric
quadrilaterals and triangles. Within each of these elements,
the hydraulic head and temperature are approximated using
biquadratic basis functions; the velocity components are
expressed by a bilinear basis function. Transient problems
are analyzed using a modified Crank-Nicholson procedure. The
actual solution of the matrix equations is carried out by a

specialized form of the Gauss elimination, developed by Irons, called the frontal solution method [4]. These procedures are fully described in References [5] and [6] since MARIAH and SAGUARO are of the same family of codes.

Due to the highly nonlinear aspects of the types of problems for which SAGUARO was designed to solve, the solution procedure contains an iterative method which allows for a specified level of convergence. At any current level of iteration the nonlinear material properties such as moisture content and permeability are obtained "implicitly" (based on latest iterated values) or "explicitly" (based on pressure values from last time step. Additionally a semi-implicit weighting can be used to obtain the material properties. A typical level of convergence is .1% based on the normalized difference in effective pressure between the present and last iterations. The number of iterations required to reach this level of convergence depends on the nonlinearity of the problem but typically can vary between two and ten. The size of the time step used has a strong influence on the number of iterations required. The time step size can be input by the user or it can be internally generated. The code generated time step is based on the geometrically increasing value. When the value gets too large to obtain convergence within the maximum allowable number of user specified iterations the code automatically reduces the time step and recalculates the solution.

6. APPLICATIONS

SAGUARO has been used to predict many aspects of the hydrological flow and energy transport regarding the design of proposed nuclear waste repositories. An example case is given here to illustrate a typical application of the code. The sample problem involves the one-dimensional infiltration of rain and heat into hard rock consisting of several strata which is initially partially saturated, see Figure 2a. The computational mesh for this problem consists of one-hundred finite elements, Figure 2b.

Non-dimensional curves, for permeability and moisture content, shown in Figure 1, are used for all strata. The dimensional value of permeability and moisture content are obtained by multiplying the non-dimensional values by these respective saturated values given in Table 1.

The initial and boundary conditions for the calculation are:
1. Rain is such that it feeds the soil at a rate of 1.0 m/yr. This rate is not typical of the Yucca Mtn. area but is used for illustrative purposes only.
2. Linear initial pressure head (ψ) at t = 0 (fluid velocity is zero everywhere). ψ (at z = 1369 m) = -620 m; ψ (at

104

z = 0) = 750 m. The capillary fringe for this case extends from the water table to the ground surface.
3. Bottom and side boundaries are considered to be imper- meable (no mass or energy flux).
4. Ground surface temperature = 35°C at t = 0 and 85°C at t > 0.

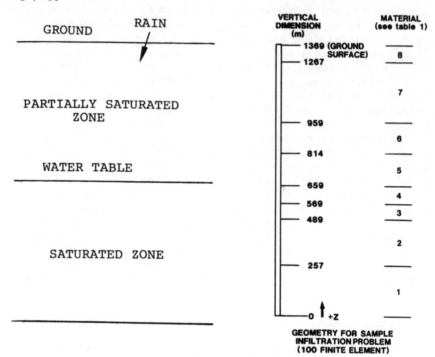

Figure 2. Geometry and Finite Element Grid Used in Sample Infiltration Problem

Strata (Fig. 2)	Material No. (Fig. 2)	Saturation Permeability $k(m^2)$	Porosity ϕ	Rock Density (kg/m^3)
1 (bottom)	2	8×10^{-16}	.32	1700
2	3	8×10^{-15}	.10	2300
3	4	8×10^{-16}	.32	1700
4	2	1.6×10^{-14}	.23	1960
5	5	1.6×10^{-16}	.24	1840
6	7	8×10^{-15}	.24	1880
7	6	6.4×10^{-14}	.20	2140
8 (top)	7	1.6×10^{-16}	.24	1840

Table 1. Material Properties

Figures 3 through 7 give the results of the sample
problem for a computed time from 0 to 10 years. The hydraulic
head, temperature and moisture are shown as a function of time
in Figures 3, 4 and 5. Figure 5 shows that the region near
the ground surface (curve 5) responds rapidly to the influx of
rain. This region is approaching saturation after ten years
of rain. In general the results show that a large response
time t (>10 yr) is required for the influence of the imposed
surface boundary conditions to propagate to depths larger than
500 m. Figure 6 gives the moisture content as a function of
depth. The soil at the ground surface is not completely
saturated at 10 years. The infiltration has not yet effected
the original location of water table. It would require on
the order of 100 years of rain to completely saturate the soil.
Figure 7 shows the vertical Darcy velocity as a function of
depth. The velocity results are in agreement with the moisture
content shown in Figure 6 in that the non-zero velocity region
and the region in which moisture is increasing correspond.

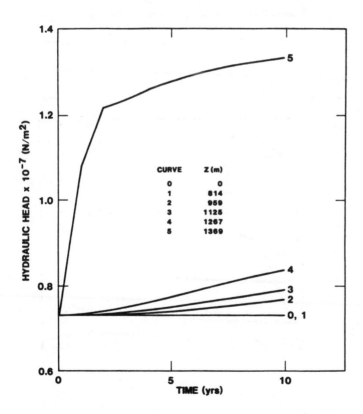

Figure 3.   Hydraulic Head as a Function of Time for Z =
            0, 814, 959, 1125, 1267, and 1369 m.

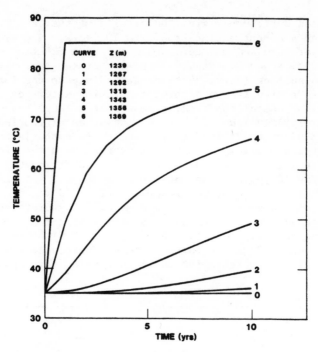

Figure 4.   Temperature as a Function of Time for Z = 1239,
1267, 1292, 1318, 1343, 1356, and 1369 m.

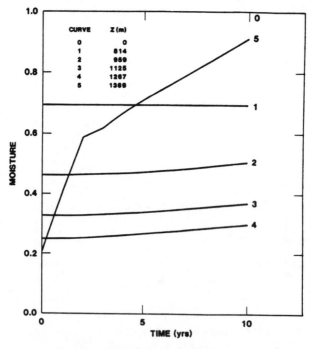

Figure 5.   Moisture Content as a Function of Time for Z = 0,
814, 959, 1125, 1267 and 1369 m.

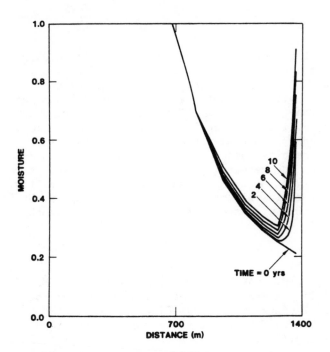

Figure 6. Moisture Content as a Function of Depth at Time =
0, 2, 4, 6, 8 and 10 yr. (Ground Surface = 1369 m)

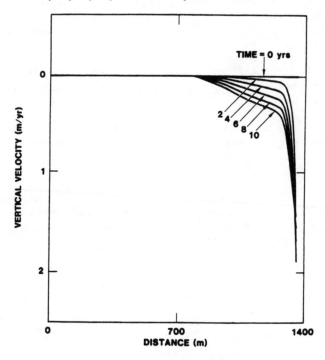

Figure 7. Velocity as a Function of Depth at Time = 0, 2,
4, 6, 8 and 10 yr. (Ground Surface = 1369 m)

108

7. CONCLUSIONS

The SAGUARO code is a user-orientated finite element code. It can be applied to a broad range of geometric configurations. The types of boundary and initial conditions that can be applied are quite general. To date applications of the code have included modeling various proposed nuclear waste repositories and associated laboratory experiments. It is concluded from these studies that the most important aspect of obtaining useful results is to be able to accurately define the material characteristic curves that relate pore pressure to moisture content and rock permeability. Additional work in this area will extend the applicability of the code to a wide range of geohydrology and geothermal problems.

8. REFERENCES

[1] EATON, R. R., GARTLING, D. K. and LARSON, D. E. - SAGUARO -- A Finite Element Computer Program for Partially Saturated Porous Flow Problems, SAND 82-2772, Sandia National Laboratories, Albuquerque, NM, March 1983.

[2] PICKENS, J. F., and GILLHAM, R. W., - Water Resources Res., 16, 1071-1078, 1980.

[3] FREEZE, R. A. and CHERRY, J. A. - Groundwater, Prentice-Hall, Inc., 1979.

[4] IRONS, B. M. - A Frontal Solution Program for Finite Element Analysis. Int. J. Num. Meth. Engng., 2, pp. 5-32, 1970.

[5] GARTLING, D. K. - MARIAH -- A Finite Element Computer Program for Incompressible Porous Flow Problems: Theoretical Background. SAND79-1622, Sandia National Laboratories, Albuquerque, NM, Sept. 1982.

[6] GARTLING, D. K. and HICKOX, C. E. - MARIAH - A Finite Element Computer Program for Incompressible Porous Flow Problems. SAND79-1623, Sandia National Laboratories, Albuquerque, NM, Aug. 1980.

CHAPTER 5

NUMERICAL PREDICTION OF BUOYANCY INDUCED FLUID FLOW IN
UNDERGROUND NUCLEAR WASTE DISPOSAL CHAMBERS

M. S. Hossain

1.   INTRODUCTION

1.1   Description of the problem

In course of safety studies for salt dome repositories
of nuclear wastes in the Federal Republic of Germany, the
consequence analysis of an accident scenario "water intrusion
into the repository" is carried out. The project "Safety
Studies Entsorgung (PSE)" has been developing a methodology
for modelling the radionuclide transport through all barriers
between wastes and man.

The present study is concerned with the barrier effect
of chambers, where only low and medium level wastes (LLW and
MLW) are planned to be disposed of. According to the present
plan [26], three different types of chambers based on available techniques will be used, depending on the type of wastes:
(1) tumble-down technique (VT) will be used for LLW-barrels,
(2) remote-stacking technique (ST) will be applied for LLW-
and MLW-barrels in lost concrete shielding (VBA), and (3) top
loading (AT) chambers will take MLW-barrels transported in
retrievable shielding. These chambers will be excavated about
850 m below surface within a salt dome repository. In this
study, the barrier effect of only the last mentioned chamber
type (AT) will be considered.

In Figure 1, the field of disposal and the arrangement
of the chambers have been shown. Figure 2 shows the detailed
geometrical configuration of one typical AT-chamber, which
has been considered in this study. The chamber is connected
with drifts at three levels, the waste disposal level (ES),
the salt excavation level (SS), and the undercut level (US).
As shown in Figure 2, after completion of disposal, these connecting drifts will be sealed with sealing material.

112

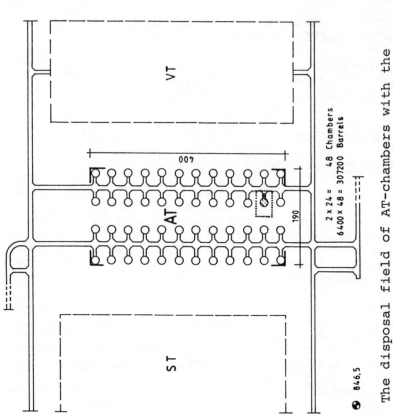

Figure 1. The disposal field of AT-chambers with the
chamber-arrangement (horizontal section at level
846.5 m below surface)

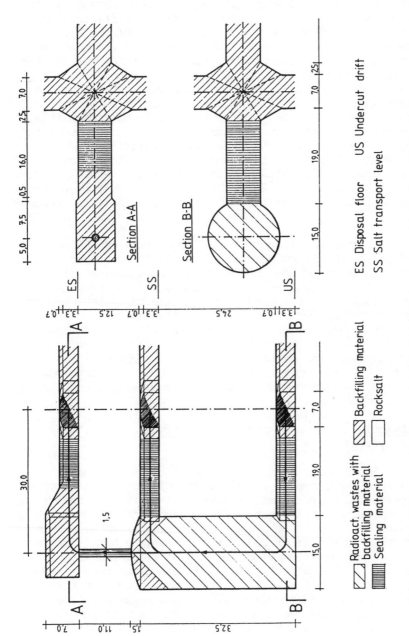

Section A-A

Section B-B

ES   US Undercut drift
SS Salt transport level

ES Disposal floor

Radioact. wastes with
backfilling material

Backfilling material

Sealing material

Rocksalt

Figure 2.   The geometrical configuration of the AT-chamber

After an accidental water-intrusion into the repository, because of high hydrostatic pressure (about 10 MPa), fluid (saturated brine) will enter into the chamber. Hereby, it is assumed that the sealings of the connecting drifts contain fractures and porous space. After saturation of the backfilled porous space and fractures in the chamber and the sealings with the fluid, the heat producing barrels inside the chamber may cause density differences in the fluid around them. These thermally induced density differences may, in turn, cause a buoyancy induced throughflow of the chamber as shown in Figure 2. This will enable the leached radionuclides within the chamber to be transported out of it. For this reason, it is important to predict the behaviour of this flow and, hence, to estimate the probable transport of the radionuclides out of the chamber. These results will be used in the general barrier model of the repository [24] which has been developed to study the effectiveness of various barrier components of the repository.

In the present situation, the flow domain can be assumed to be a fluid saturated porous medium. Basic equations governing the transport of heat and mass in porous medium exist. These are the continuity equation, the momentum equation (Darcy's law) and the energy equation. The main problem is to solve these equations in complex geometric situations for obtaining the distribution of required quantities, e.g. velocity, temperature, etc. within the flow domain. The complexity in practical situations is caused by (1) the shape of the volume containing the porous medium, (2) properties of the porous medium and its surrounding, and (3) hydrodynamical and thermal boundary conditions.

1.2 Existing methods of predicting thermal convective flows in porous media

The availabe methods of prediction can be broadly classified into experimental and mathematical. Experimental methods are restricted only to small geometrical configurations of the laboratory with simple boundary conditions and are suitable to study the qualitative behaviour of the flow or to determine some global parameter of the flow like average heat transfer. Thus, for the quantitative prediction of practical flow situations, one is left with mathematical methods which are based on the basic equations of flow. For some simple flow situations with idealized boundary conditions, exact solutions of the basic equations can be obtained. Thus, Horton and Rogers [16], Lapwood [21] and Nield [23] obtained such analytical solutions towards the criterion for the onset of convective motion in porous media, which are also well verified by experiments (e.g. Schneider [27], Elder [14], Katto and Masouka [19], Combarnous and Le Fur [12]. The review article of Combarnous and Bories [11] and Cheng [8] report these in details. Aziz and Combarnous [1] and Bories [4] obtained also

an analytical solution concerning the mean heat transfer in the
case of horizontal layers bounded by isothermal planes which
agrees satisfactorily with the experimental results [5, 12, 14,
18, 27, 29]. But such analytical results are too rough to be
applied to practical situations, where more precise informa-
tions are required.

Thus, numerical methods of different complexity are being
developed in which the geometry and the boundary conditions
can be incorporated more realistically. Numerical models for
the simple case of a two-dimensional (2-D) roll have been developed by many authors (e.g. Aziz et al. [2], Combarnous [9]
and Vlasuk [28]). The results of these models compare well with
the experiments mentioned above. More sophisticated models to
incorporate the influence of temperature difference between
the fluid and solid phases (Combarnous [10], Chan and Banerjee
[7]) or to include the effect of inertia in the Darcy's law
(Chan and Banerjee [7]) are being developed for special flow
situations.

Numerical models for three-dimensional geometry are also
available. The 3-D-model of Chan and Banerjee [7] is applicable to simpler geometries and homogeneous medium. The numerical scheme of INTERA [17], developed for groundwater flow calculations, is more general and applicable to heterogeneous
medium in geometries varying form 1-D to 3-D.

1.3 Present contribution

The foregoing discussion has shown that only numerical
methods can be applied to predict realistically a practical
flow situation. Thus, efforts have been devoted to predict the
flows in the 2-D-model geometry of the AT-chamber described in
section 1.1 using the numerical scheme developed by INTERA.

In the following section, the basic equations governing
the flow in porous media are introduced. The simplifying assumptions are discussed briefly. The 3-D numerical scheme of
INTERA (SWIFT: Simulator for Waste Injection, Flow and Transport) towards the solution of these equations is then described in short.

Subsequently in Section 3, the capability of SWIFT to
predict free convection flows (Rayleigh-convection) in 2-D-
geometries has been tested, for which experimental results
are available.

Section 4 describes the modelled geometrical configuration of the AT-chamber used in the present 2-D-calculations.
The properties of the materials in different regions of flow
domain have been estimated, and the boundary and initial conditions have been discussed. Section 5 concerns the application of the SWIFT program in the 2-D-model geometry of the

chamber described in Section 4. The obtained results are discussed and a parameter study is carried out. Finally, in Section 6 the obtained results are summarized and concluding remarks have been drawn.

2. BASIC EQUATIONS OF FLOW AND THE METHOD OF SOLUTION

2.1 Basic equations

The laminar flow of a single fluid (variation of density is caused by temperature only) flowing through a porous matrix obeys the laws of conservation of mass, momentum and thermal energy which, when expressed in mathematical form, yield the time-dependent continuity, momentum and energy (for both fluid and solid matrix) equations. Together with an equation of state, these equations form a closed set that describes all the details of the motion inside the pore space (microscopic scale). This motion cannot be resolved for practical problems with present day computers in a numerical solution, so that a passage from the microscopic scale to the macroscopic one is carried out by an averaging procedure, accompanied by some conceptual model of the porous medium. This procedure has been discussed in more detail by Dagan [13]. The result of such a procedure is a macroscopic mathematical description of the problem of fluid flow in a porous medium, (Bear [3]) which, after further simplification according to an order of magnitude analysis in the momentum and energy equations following Dagan [13], is a closed set of equations as follows:

Continuity equation:

$$- \nabla \cdot (\rho \vec{u}) = \frac{\partial}{\partial t} (\phi \rho_f) \tag{1}$$

Momentum equation (Darcy's law):

$$\vec{u} = - \frac{k}{\mu} (\nabla p - \rho_f g \nabla z) \tag{2}$$

Energy equation:

$$\underbrace{- \nabla \cdot (\rho_f c_{pf} T \vec{u})}_{\text{convection}} + \underbrace{\nabla \cdot (\lambda_e \cdot \nabla T)}_{\text{diffusion}} \underbrace{- q}_{\text{source}} = \underbrace{\frac{\partial}{\partial t} \left[\phi \rho_f c_{pf} T + (1-\phi) \rho_s c_{ps} T \right]}_{\text{rate of change}} \tag{3}$$

Equation of state:

$$\rho_f = \rho_{fo} \left[1 - \beta (T - T_o) \right] \tag{4}$$

where $\vec{u}$ is the velocity vector representing Darcy-velocity, T the temperature, p is the pressure, ρ_f the fluid density with a reference value ρ_{fo} at temperature T_o, ρ_s the density of the solid medium within the porous matrix, g the component of gravitational acceleration in z-direction (vertical), μ and λ_e are the molecular viscosity of fluid and the effective thermal conductivity of the porous medium respectively, c_{pf} and c_{ps} are specific heat at constant pressure for fluid and solid respectively, β is the coefficient of thermal expansion of fluid, φ and k are porosity and permeability of the porous medium and q represents the heat source within the medium.

The term containing g in the Darcy's law (2) represents the influence of gravitational force and accounts for buoyancy effects in situations of non-uniform density. It should be emphasized that Boussinesq-approximation has not been introduced into the Equations (1) to (3) so that density variations are accounted for also in the continuity equation (1) and in the convective terms of the energy equation (3), and not only in the gravitational term of the Darcy-equation (2). Hence, the equations are applicable in situations with relatively large density differences.

In the momentum equation (2), the inertial terms have been neglected through an order of magnitude analysis following Dagan [13], the flow being at small Reynolds number. Strictly speaking, the diffusion term in the energy equation (3) should contain a dispersive part, but it is negligible compared to thermal conduction in the Darcian flow regime (Green [15]). Further, the description of heat transfer by the energy equation (3) assumes that in an elementary volume the mean temperatures of the solid and fluid phases are equal and an effective thermal conductivity λ_e is valid which can be expected in natural porous media saturated with water. In general, permeability k is a tensor. In this work, an isotropic porous medium is considered for which k is a scalar quantity.

2.2 Method of solution

The basic equations of flow introduced in Section 2.1 represent a complete system for the quantities u, ρ, T and p. There are, however, no simple means of reducing it by elimination because of complex coupling of parameters in these equations. Thus, a numerical procedure is needed to solve these equations.

In the present work, the numerical scheme of INTERA (SWIFT) [17] is adopted. The main features of this scheme are shown in Figure 3. Based on the mathematical model described in Section 2.1, the calculation of a particular case characterized by the geometry and material properties can be carried out, when the initial and boundary conditions for the required quantities like velocity and temperature are supplied. In the computer code, the partial differential equations of flow are

converted to time- and space-discretisized finite-difference
equations. These equations are then solved with the help of a
solution method to obtain the distribution of the required
quantities in the flow domain.

In the SWIFT-code, a choice of geometry varying from 1-D
to 3-D is possible. Inside the geometry, the material properties can be varied. The code offers a wide choice of boundary
conditions. For the finite difference approximations, the user
has a choice of difference schemes varying from central-in-
space-and-time to backward-in-space-and-time to cope with the
problem of numerical dispersion and stability of solution. For
the solution procedure itself, there are two choices: direct
solution and iterative method. The calculations of the present work were carried out using a central-in-time and back-
ward-in-space difference scheme with the direct method of
solution.

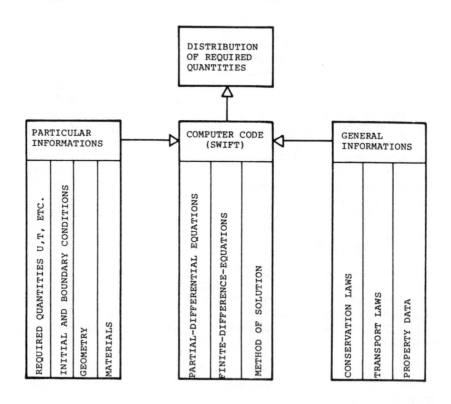

Figure 3:   The main features of the SWIFT-code

3.   APPLICATION OF SWIFT TO PREDICT FREE CONVECTION FLOWS
     IN POROUS MEDIUM

     The mathematical model described in Section 2 will now
be applied to predict free convection flows in porous medium.
The purpose of this application is to check the capability of
the SWIFT-code to predict free convection flows, for which the
code has not been tested before. The flow considered is the
Rayleigh–Bénard–convection [25] in porous medium of two-
dimensional geometry, for which an analytical solution can be
obtained.

3.1   Description of the flow

     A schematic sketch of free convection flow in porous
medium is given in Figure 4. The homogeneous and isotropic
porous layer saturated with fluid is bounded by isothermal
impervious horizontal planes distance H apart, and it is heated from below $(T_B > T_T)$. In experimental configurations, the
side walls are kept insulated, so that heat is transported
only in the vertically upward direction. It has been observed
in experiments that up to a certain difference of temperature
$(T_B - T_T)$ the fluid medium remains at rest and the heat transfer takes place in the conductive mode alone. At a particular
temperature difference, cellular movements in the fluid medium
start. Due to these movements, heat is now transported in the
combined conductive and convective mode. The important parameters governing this flow will now be described.

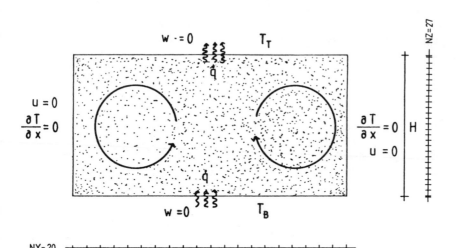

Figure 4.   The geometry and boundary conditions
            for the calculation of free convection
            flow

3.2 Parameters of the flow

Nondimensionalization of the equations in Section 2.1 leads to an important parameter called Rayleigh number:

$$Ra = \frac{g \beta k \, (T_B - T_T) \, H}{(\mu/\rho_f) (\lambda_e/\rho_f c_{pf})}, \tag{5}$$

which characterizes the relative magnitudes of the buoyant and resisting forces. As discussed in Section 3.1, at some critical value of Rayleigh number, depending on bed geometry, boundary conditions and fluid and bed properties, the energy transport mechanism changes from conduction to convection. The ratio of the energy transport by conduction and convection is given by the layer Nusselt number defined by

$$Nu = \frac{\dot{q}}{\lambda_e (T_B - T_T)/H} . \tag{6}$$

Here, $\dot{q}$ is the rate of energy transfer per unit cross-sectional area of the bed.

3.3 Analytical solution

For the flow configuration described analytical solutions are obtainable towards the criterion for the onset of free convection and to compute the mean heat transfer caused by convection.

(a) Criterion for the onset of free convection

Conditions for the occurence of free convection are obtained by using a small disturbances method. Hereby a linear theory is applied which assumes that all products and powers higher than the first of the disturbance components appearing in the equations are negligible. Starting from an initial state corresponding to a motionsless fluid, the critical Rayleigh number for the appearance of the first convective mode (see [11] for detailed derivation) is:

$$Ra_c = 4 \pi^2 . \tag{7}$$

(b) Mean heat transfer

The main consequence of convective motion is to increase the mean heat transfer expressed by the dimensionless Nusselt number Nu defined by Equation (6). This increase in the mean heat transfer due to convection cannot however be predicted

using the linear theory, because any product of disturbance
components is omitted by the basic assumption. Using the power
integral method developed by Malkus and Veronis [22] in case
of fluid volumes, both Aziz and Combarnous [1] and Bories [4]
obtained an analytical relationship for the overall heat transfer. For a Rayleigh number range (60 - 4000), this relation
can be approximated as

$$Nu = 0.218 \ Ra^{0.5}. \tag{8}$$

3.4 Experimental data

In the case of horizontal layers bounded by isothermal
planes, a lot of experimental results is now available. For
the convective steady states, the main results concern the
criterion for the onset of convection and the mean heat transfer. The standard criterion for the onset of free convection
given by Equation (7), Ra ≃ 40, is confirmed by experimental
results (Figure 6). If Ra is higher than 40, but not too large,
a stable convective state exists which is characterized by
adjacent polyhedral cells of two-dimensional rolls. Another
convective state was found for Ra higher than a critical value
lying between 240 - 280 depending on the porous medium [11].
This state corresponds to a relative increase of heat transfer
compared to the state mentioned above and is characterized by
a continuously fluctuating temperature distribution inside the
porous layer. For this reason, it has been called [11] fluctuating convective state.

An extensive set of results is available concerning the
mean heat transfer Nu in the steady state (Figure 6). Ra
varies from 0 to 2000. The scatter of the relationship between
Nu and Ra of the experiments could be due to its dependance on
the thermal characteristics of the medium. The definitions of
Rayleigh and Nusselt numbers in Equations (5) and (6) assume
an effective thermal conductivity of the porous medium λ_e.
This is only true when the heat transfer coefficient between
the solid and fluid phases is infinite; but this is not the
case in experimental situations.

3.5 Results of calculation and comparison with experiments

Using the thermal and hydrodynamic boundary conditions as
shown in Figure 4, numerical calculations with the help of
SWIFT-code were carried out. The steady state temperature and
velocity distributions calculated for a Rayleigh number
Ra > Ra_c are shown in Figure 5. The two-dimensional rolls are
distinctly visible, causing a corresponding temperature distribution. Heat transfer takes place both in the conductive
and convective mode.

122

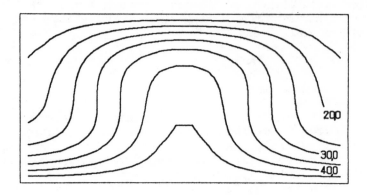

(a) Temperature distribution

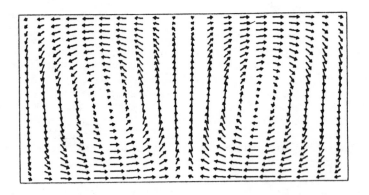

(b) Velocity distribution

Figure 5.   The calculated temperature and
velocity distribution for free
convection flow in porous medium
(Ra ≃ 77)

123

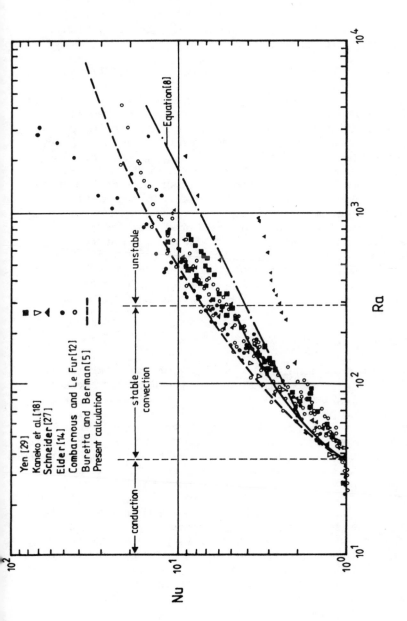

Figure 6.   The relation between Nu and Ra

In Figure 6, the calculated relationship between Nu and Ra has been compared with the analytical solution (Equation (8)) and experiments. Considering the scatter of the experiments for possible reasons discussed in Section 3.4, the calculated bold curve compares farely well with the experiments and the analytical solution. Steady state solutions in the calculations could however be obtained only upto Ra ≈ 200. Beyond that, the temperature fluctuations were very high causing instability in the numerical solution.

4. SIMPLIFIED MODELLING OF THE PRESENT FLOW SITUATION

The actual configuration of the chamber, where the barrels containing radioactive wastes will be deposited, has already been shown in Figure 2. The material constituting the porous medium inside the geometry is heterogeneous due to different ingredients present in different zones of it. For the numerical calculation with a finite difference method (SWIFT), space-discretization in the form of rectangular mesh must be carried out. Hence, some simplification of the geometry is necessary and the material properties of the heterogeneous porous medium should also be approximated. Heat sources in the disposal area should also be known. Further, to start the time dependent calculations, initial values of the variables at t = 0, and conditions at the boundaries of the calculation during the run must be specified. Some simplifying assumptions are also necessary for that. These have been described below in details.

4.1 Geometry

The two-dimensional model geometry of the chamber used in the calculation is shown in Figure 7, the dimension perpendicular to this plane (in y-direction) being unity. This geometry can be compared with the vertical longitudinal section through the chamber shown in Figure 2, the curvatures and inclinations being evenned out. As shown in Figure 7, some part of the rock-salt (designated as ③) has to be included in the calculation, to be able to assign proper boundary conditions. The adopted space-discretizations in both the horizontal (NX) and vertical directions (NZ) have also been shown in it. In zone ①, the barrels containing radioactive wastes will be deposited, the rest volume being backfilled with excavated salt. In zones ④, sealing material is present, whereas zone ② contains only backfill material. The narrow vertical slit of the chamber, designated by zone ⑤, contains sealing material. The estimation of the properties in these different zones will now be carried out in the next section.

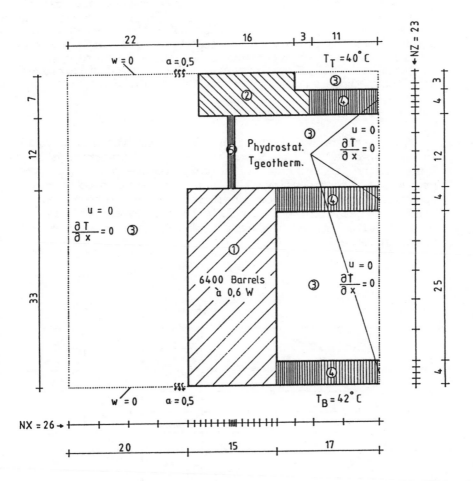

Figure 7.   The 2-D model geometry of the
            chamber with boundary conditions
            for calculation

4.2 Material properties

As described in Section 2, for the calculation of the flow, the following properties of the fluid (brine) saturating the porous matrix must be known: density ρ_f, coefficient of thermal expansion β, specific heat at constant pressure c_{pf} and viscosity μ. For the saturated porous matrix, the effective values of porosity ϕ, permeability k and thermal conductivity λ_e should also be estimated. Further, for the time dependent calculations the density of the solid material in the porous matrix ρ_s and its specific heat at constant pressure c_{ps} should also be supplied; however, in the aspired steady state solutions, these values do not have any influence.

(a) Properties of the fluid

It is assumed that the fluid saturating the porous matrix is saturated NaCl-brine. Since the ranges of temperature difference in the present cases of calculation are small, all other properties except density ρ_f of brine are assumed to be constant. The variation of density is assumed to be linear, i.e. the coefficient of volumetric thermal expansion β is constant. The thus adopted values of the fluid properties are given in Table 1.

Zone		ϕ	k	λ_e
		$-$	m^2	$W m^{-1} K^{-1}$
①	Rocksalt-concrete-water	0,2	$5 \cdot 10^{-11}$	1,5
②	Rocksalt-water	0,4	10^{-10}	2,3
③	Rocksalt	0,001	10^{-18}	5,0
④	Rocksalt-water	0,2	10^{-11}	3,5
⑤	Rocksalt-water	0,2	$5 \cdot 10^{-12}$	3,5

ρ_{fo}	β	c_{pf}	μ	λ_f	ρ_s	$\rho_s c_{ps}$
$kg\,m^{-3}$	K^{-1}	$J kg^{-1} K^{-1}$	$Pa\,s$	$W m^{-1} K^{-1}$	$kg\,m^{-3}$	$J m^{-3} K^{-1}$
1194,0 $(T_0 = 40°C)$	$3,1 \cdot 10^{-4}$	4000	$1,5 \cdot 10^{-3}$	0,5	2100	$1,9 \cdot 10^6$

Table 1. The properties of material used for the reference case of calculation

(b) Properties of the porous matrix

While defining porosity of a porous medium, distinction is often made between (1) the absolute or total porosity, defined as the ratio of the total pore volume to the total volume of the medium regardless of whether pores are interconnected or not, and (2) the effective porosity, whereby only the interconnected (effective) pores are considered. The latter, pertaining to the interconnected pores only, is of interest from the standpoint of flow through the porous medium; thus, by porosity ϕ, in this work, is meant the effective porosity. The porosities in different zones, used for the present calculations, are given in Table 1. The backfill material in zone ② is assumed to have a porosity of 0.4, whereas the sealing material in zones ④ and ⑤ 0.2 . In zone ①, where barrels occupy about 50 per cent of the volume and the rest is backfill material, the porosity has been reduced to 0.2. For rocksalt (zone ③), the commonly used porosity of 0.001 has been adopted.

Permeability is the most important property governing the flow in porous medium. Since it originates from the constant of proportionality in Darcy's law, usually called hydraulic conductivity K (in older text, also referred as the coefficient of permeability), the two definitions are often confused. To avoid this, the relation between the two definitions is given below:

$$k \atop [m^2] = \frac{\mu}{\rho_f g} \cdot K \atop [m \cdot s^{-1}] \qquad , \qquad (9)$$

with their units put between parenthesis underneath. The permeability k is a function only of the medium, and has the dimension m^2; thus, the existence of gas, water or any other fluid in the porous medium does not change its value. In the absence of enough reliable measured data for permeability of materials (both backfill and sealing) present in different zones of the flow geometry in Figure 7, some conservative guess has been made for the reference case, as shown in Table 1. For the porous materials in zones ①, ②, ④ and ⑤ the permeability varies between 10^{-10} to $5 \cdot 10^{-12}$ m^2. In the disposal zone ①, where waste barrels with backfill material are present, a composite permeability of the combination is estimated using the series arrangement assumption after Bear [3]. Thus the permeablity in zone ① is half of that in zone ②, where only backfill material is present. The permeability in zone ④ has been taken one order-of-magnitude smaller than in ②. The permeability in zone ⑤ of the chamber has been further reduced, since practical experience has shown that sealing material in such vertical boreholes can be compacted more effectively.

To describe the transmission of heat in a porous medium, an equivalence is often used between the heterogeneous porous medium, made up of a solid matrix and saturating fluids, and a fictitious continuum, in which is defined a heat transfer equation similar to the equation used, for instance, in homogeneous fluids. This equivalence which employs an equivalent thermal conductivity λ_e is justified when the saturating fluid velocities are not too high and if the average grain size of the porous medium is small. Concerning the values of effective thermal conductivity λ_e, compared with the conductivities of the solid and fluid phases λ_s and λ_f, numerous models are available. For the estimation of the effective thermal conductivities in different zones of the present case, a method suggested by Krischer [20] has been used. These values are given in Table 1.

4.3 Rate of heat generation

The heat producing wastes, contained in barrels, are deposited in zone ① of the chamber. The rate of heat production q_s per barrel, which had already been estimated in a previous study [24], has been used in this calculation. For the reference case of calculation, the rate of heat generation is assumed to be 0.6 W per barrel, which is the estimated rate for the type of radionuclides contained in the barrels after a cooling period of ten years. The total amount of heat has been distributed uniformly in the disposal zone ①.

4.4 Boundary and initial conditions

As shown in Figure 7, the no-flow boundary conditions have been used across all the sides, except the connecting drifts (Zone ④) on the right side. Across these drifts, it is assumed that the pressure distribution is hydrostatic, which would mean that these drifts are connected with an infinite reservoir.

As regards the thermal boundary conditions, on the upper and lower side of the chamber the "radiation" type boundary condition of Carslaw and Jaeger [6] (Type D) has been used. This specifies a linear heat transfer at the surface given by

$$F = \alpha (T - T_o), \tag{10}$$

where F is the heat flux, α the coefficient of surface heat transfer, T the temperature of the surface and T_o that of the surrounding media. The value for the coefficient of heat transfer α, as given in Figure 7, is obtained assuming that at about 10 m above the top surface and below the bottom surface respectively the undisturbed geothermal temperature is present and the heat transfer inbetween takes place in vertical conduction mode only. The surrounding temperatures T_T and T_B

used in the present case have been given in Figure 7. Since
both on the left and right sides, the symmetry condition
caused by the neighbouring chambers is present, $\partial T/\partial x = 0$,
except across the connecting drifts, where geothermal temperatures have been specified.

The initial conditions for the present calculation case
are not very important, since only the steady state solutions
are aspired. However, the constant mean geothermal temperature
in the chamber area and a hydrostatic pressure distribution
have been adopted for the initial temperature and pressure
respectively.

5. CALCULATION OF THE SIMPLIFIED FLOW SITUATION AND
    DISCUSSION OF RESULTS

In this section, the results of calculation with the help
of SWIFT-code for the simplified two-dimensional flow situation of the chamber described in Section 4 are presented and
discussed. At first, the reference case has been calculated;
then, calculations were carried out with parameter variations,
to study the influence of the two important parameters, permeability and heat production rate, on the throughflow of the
chamber.

5.1 Velocity and temperature distribution

The velocity and temperature distribution for the
reference case of the chamber are shown in Figure 8. In Figure 8(a), the calculated temperature field has been shown.
The rise of temperature is maximum around the top of the disposal chamber which is about 18 K. This rise of temperature
causes a rise across the edge of the middle connecting drift
(SS) of about 10 K. Thus, the values for pressure and temperature across the middle drift, which are supplied as boundary conditions, have to be corrected iteratively. The velocity field (Figure 8(b)) shows that there is a strong throughflow of the disposal zone ①. The most of the throughflow
takes place through the lower sealing (inward), disposal zone
(upward), and the middle sealing (outward). It can also be
seen that there are two very weak counter rotating cells
within the disposal area causing downward flow along the walls.
A small part of the fluid is forced up through the vertical
slit (zone ⑤), which has less permeability than the horizontal sealings. This volume of fluid (about one-fourth of the
total) is finally forced out through the upper sealing. The
flow in the disposal floor (zone ②) is characterized by two
strong counter rotating cells, since the permeability here is
comparatively higher.

130

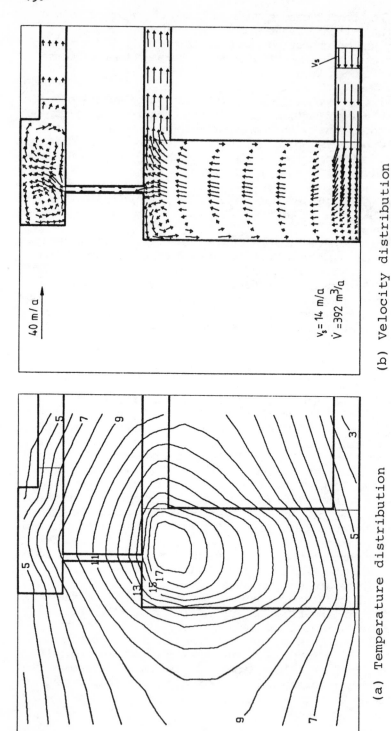

40 m/a

v_s = 14 m/a
$\dot{V}$ = 392 m³/a

v_s

(a) Temperature distribution   (b) Velocity distribution

Figure 8.   The calculated temperature and velocity distribution for the reference
             case of the chamber

From the velocity through the lower sealing v_s the rate of throughflow of the chamber $\dot{V}$ can be calculated (Figure 8(b)) as

$$\dot{V} = v_s \cdot A, \tag{11}$$

where A is the area of cross-section of the sealing drifts.

5.2 Parameter-study

As mentioned in Section 4.2, enough reliable measured data of permeability of backfill and sealing materials are not yet available. Thus, the used values for the reference case are only guesses. Since it is an important parameter influencing the throughflow, and since it is expected that, due to creep of the surrounding rock salt, the permeabilities within the chamber will be reduced with time, it is important to know how it affects the throughflow. Another important parameter of flow is the rate of heat generation, which is known to decrease with time [24]. Thus, in this section, the influence of these two parameters on the throughflow of the chambers have been investigated.

(a) Permeability

At first, a qualitative influence of the reduction of permeability of both the backfill and sealing materials on the throughflow has been studied. For that the permeabilities of both the backfill and sealing materials have been reduced separately by one order of magnitude.

Figure 9 shows the influence of the reduction of permeability on the throughflow of the chamber. In the middle (Figure 9 (b)), the reference case of Figure 8 has been shown. Above, in Figure 9(a), the permeability of the sealing material (zones ④ and ⑤) has been reduced, which is shown to have little influence on the flow in the disposal zone, whereas the throughflow has been drastically reduced. Below, in Figure 9(c), the permeability of the backfill material (zones ① and ②) has been reduced, which reduces the cellular motion in the disposal zone, whereas the reduction of the flow through the sealings is negligible.

A quantitative presentation of these influences has been done in Figure 10. The throughflow decreases almost linearly with the decrease of sealing permeability, whereas the permeability of the disposal zone has little effect on the throughflow.

132

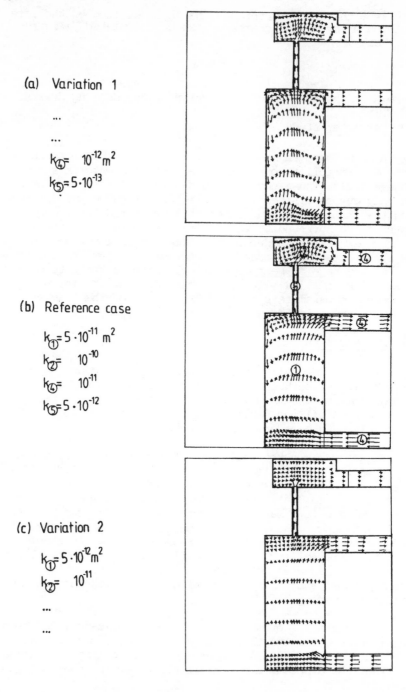

(a) Variation 1

...

...

$k_{\textcircled{4}} = 10^{-12} m^2$
$k_{\textcircled{5}} = 5 \cdot 10^{-13}$

(b) Reference case

$k_{\textcircled{1}} = 5 \cdot 10^{-11} m^2$
$k_{\textcircled{2}} = 10^{-10}$
$k_{\textcircled{4}} = 10^{-11}$
$k_{\textcircled{5}} = 5 \cdot 10^{-12}$

(c) Variation 2

$k_{\textcircled{1}} = 5 \cdot 10^{-12} m^2$
$k_{\textcircled{2}} = 10^{-11}$

...

...

Figure 9. The influence of permeability on the throughflow of the chamber

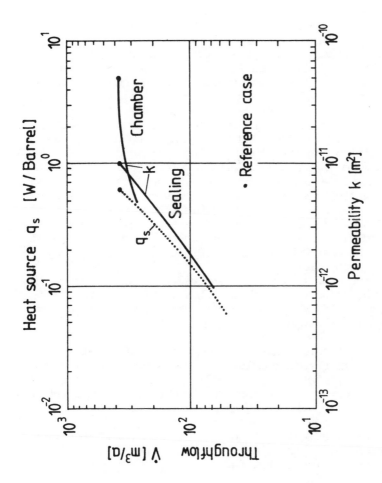

Figure 10. Influence of permeability and rate of heat generation on the throughflow of the chamber

(b) Rate of heat production

The rate of heat generation from the wastes decreases with time. Thus, using several heat production rates, which are expected at different times, steady state calculations were carried out to study the influence of this very important parameter on the flow. The result of this study has been shown in Figure 10; with the decrease of heat production, the troughflow of the sealings decreases almost linearly.

6.   CONCLUDING REMARKS

In this study, a numerical prediction of the throughflow of underground chambers caused by the thermal convection due to heat generation has been presented. Assuming that the medium is a fluid saturated porous medium, the solution of the basic equations of flow has been done to obtain the velocity and temperature distributions in the flow domain. The geometrical configuration of the flow and the boundary conditions have been simplified, so that the numerical scheme of INTERA, SWIFT-code, could be efficiently applied. Before applying the SWIFT-code to calculate the chamber flow, its applicability has been tested by applying to a simple free convection flow case, for which experimental data are available. Then, calculation of the reference case has been done for the assumed model geometry of the chamber with assumed heterogeneous properties of material in it and the assumed boundary conditions.

The predicted temperature and velocity distributions for the reference case of the chamber have been presented. The rates of throughflow of the chamber have been obtained from the velocities through the sealings.

The investigation has been concluded with a parameter study. As an important conclusion, it has been shown in this parameter study that the permeability of the sealings and the rate of heat generation within the chambers are important parameters influencing the throughflow of the chamber.

The numerical calculations reported in this study have been carried out in the CYBER 170-835 computer of the Technical University of Berlin.

REFERENCES

1.   AZIZ, K. and COMBARNOUS, M. A. - Prédiction théorique du transfert de chaleur par convection naturelle dans une couche poreuse horizontale, C. R. Acad. Sci., Ser. B 271, pp. 813-815, 1970.

2.   AZIZ, K., HOLST, P. H. and KARRA, P. S. - Natural Convection in Porous Media, Petrol. Soc. Can. Inst. Mining, Calgary, Pap. 6813, 1968.

3. BEAR, J. - Dynamics of Fluid in Porous Media, American Elsevier Pub. Co., New York, 1972.

4. BORIES, S. - Comparison des prévisions d'une théorie non linéare et des resultats experimentaux en convection naturelle dans une couche poreuse saturée horizontale, C.R. Acad. Sci., Ser. B271, pp. 269 - 272, 1970.

5. BURETTA, R.J. and BERMANN, A.S. - Convective Heat Transfer in a Liquid Saturated Porous Layer, Trans. ASME, J. Appl. Mech., Vol. 98, pp. 249 - 253, 1976.

6. CARSLAW, H.S. and JAEGER, J. C. - Conduction of Heat in Solids, Oxford Univ. Press, London, 1959.

7. CHAN, Y. T. and BANERJEE, S. - Analysis of Transient Three-Dimensional Natural Convection in Porous Media, Trans. ASME, J. Heat Transfer, Vol. 103, pp. 242 - 248, 1981.

8. CHENG, P. - Heat Transfer in Geothermal Systems, Advances in Heat Transfer, Vol. 14, pp. 1 - 105, 1978.

9. COMBARNOUS, M. - Etude numerique de la convection naturelle dans une couche poreuse horizontale, C.R. Acad. Sci., Ser. B271, pp. 357 - 360, 1970.

10. COMBARNOUS, M. - Description du transfert de chaleur par convection naturelle dans une couche poreuse horizontale à l'aide d'un coefficient de transfert solide-fluide, C.R. Acad. Sci., Ser. A 275, pp. 1375 - 1378, 1972.

11. CAMBARNOUS, M.A. and BORIES, A.A. - Hydrothermal Convection in saturated porous media, Advances in Hydroscience, Vol. 10, pp. 231 - 307, 1975.

12. COMBARNOUS, M. and LE FUR, B. - Transfert de chaleur par convection naturelle dans une couche poreuse horizontale, C.R. Acad. Sci., Ser. B 269, pp. 1009 - 1012, 1969.

13. DAGAN, G. - Some Aspects of Heat and Mass Transfer in Porous Media, IAHR Symp. Fundamentals of Transport Phenomena in Porous Media, Haifa, 1968.

14. ELDER, J. W. - Steady Free Convection in a Porous Medium Heated from Below, J. Fluid Mech., Vol.27, pp.29-48, 1967.

15. GREEN, D. W. - Heat Transfer with a Flowing Fluid Through Porous Media, Ph. D. thesis, Univ. of Oklahoma, 1963.

16. HORTON, C. W. and ROGERS, F. T. - Convection Currents in a Porous Medium, J. Appl. Phys., Vol.16, pp. 367-370, 1945.

136

17. INTERA ENVIRONMENTAL CONSULTANTS - The INTERA Simulator for Waste Injection, Flow and Transport - User's Manual for SWIFT, Version TUB/PTC, Release 3.82, INTERA Environmental Consultants, Inc., Houston, 1982.

18. KANEKO, T., MOHTADI, M. F. and AZIZ, K. - An Experimental Study of Natural Convection in Inclined Porous Media, Int. J. Heat Mass Transfer, Vol. 17, pp. 485 - 496, 1974.

19. KATTO, Y. and MASOUKA, T. - Criterion for Onset of Convective Flow in a Fluid in a Porous Medium, Int. J. Heat Mass Transfer, Vol. 10, pp. 297 - 309, 1967.

20. KRISCHER, O. - Die wissenschaftlichen Grundlagen der Trocknungstechnik, Springer-Verlag, Berlin, 1963.

21. LAPWOOD, E. R. - Convection of a Fluid in a Porous Medium, Proc. Cambridge Phil. Soc., Vol. 44, pp. 508 - 521, 1948.

22. MALKUS, W. V. R. and VERONIS, G. - Finite Amplitude Cellular Convection, J. Fluid Mech., Vol. 4, pp. 225 - 260, 1958.

23. NIELD, D. A. - Onset of Thermohaline Convection in a Porous Medium, Water Resources Res., Vol. 4, No. 3, pp. 553 - 560, 1968.

24. PSE - Zusammenfassender Zwischenbericht, Projekt Sicherheitsstudien Entsorgung, Berlin, 1981.

25. RALEIGH, LORD - On the Convection Currents in a Horizontal Layer of Fluid When the Higher Temperature is on the Underside, Phil. mag., Vol. 32, pp. 529 - 546, 1916.

26. RÖTHEMEYER, H. - Site Investigation and Conceptual Design for the Mined Repository in the nuclear "Entsorgungszentrum" of the Federal Republic of Germany, Underground Disposal of Radioactive Wastes, Vol. 1, PAPER IAEA-SM-243/48, IAEA, Vienna, 1980.

27. SCHNEIDER, K. J. - Die Wärmeleitfähigkeit körniger Stoffe und ihre Beeinflussung durch freie Konvektion, Dissertation, Univ. of Karlsruhe, 1963.

28. VLASUK, M. P. - Transfert de chaleur par convection dans une couche poreuse (in Russian), All-Union Heat Mass Transfer conf., 4th, Minsk, 1972.

29. YEN, Y. C. - Effect of Density Inversion on Free Convective Heat Transfer in Porous Layer Heated from Below, Int. J. Heat Mass Transfer, Vol. 17, pp. 1349 - 1356, 1974.

CHAPTER 6

THE BÉNARD PROBLEM REVISITED

J. RAE

Theory of Fluids Group, Harwell, England

1.   INTRODUCTION

The Rayleigh-Bénard problem of the motion of the fluid
in a rectangular cavity that is uniformly heated from below
is of practical and theoretical importance. It has applications in such diverse branches of physics as meteorology and
astrophysics, and provides a simple example of bifurcations
in a fluid flow problem.

The physical phenomena can be summarised as follows. If
the Rayleigh number Ra (which is a dimensionless measure of
the temperature difference across the cavity) is less than a
certain critical value Ra_{crit} then there is no flow and a
uniform vertical temperature gradient. If Ra is greater than
Ra_{crit} then the no flow state is unstable, and steady convection cells form. Ra_{crit} and the number of convection cells
that form depend upon the width-to-height ratio of the cavity,
and upon the boundary conditions upon the walls of the cavity.
The original studies [1,2,3] were for the classical case of a
two-dimensional infinitely wide cavity for which the critical
Rayleigh number is 1708. It is sometimes possible to realise
these two-dimensional flows experimentally although sometimes
three-dimensional effects are crucial. Here we restrict ourselves to the 2-D cases.

At the critical Rayleigh number there is a bifurcation
of the solution of the non-linear steady flow equations. The
lowest such bifurcation occurs as in Figure 1.

If Ra is only slightly greater than Ra_{crit} then the flow
velocities are small, and so Ra_{crit} can be obtained from a
linear eigenvalue problem derived by neglecting the non-linear
terms in the steady flow equations. However, only a few
choices of boundary conditions can be solved analytically.
In particular the case of a finite cavity with rigid walls on

all sides requires numerical treatment. A numerical approach
is also necessary to obtain the magnitude of the flow for
values of Ra greater than Ra$_{crit}$ where the non-linear terms
matter.

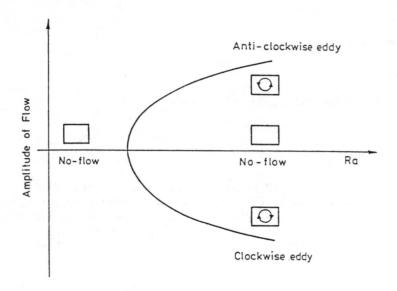

Fig. 1   Schematic diagram of the bifurcation in the
          Bénard problem at Ra$_{crit}$. The signed amplitude of
          the flow is plotted against Ra. (The flow on the
          various branches is indicated by the pictures).

We use here a standard finite-element program for
buoyancy driven flow to examine the critical parameters and
relations among the possible flows for the square cavity, the
tilted cavity and for cavities with aspect ratios greater
than 1. The numerical results reveal a rich structure of
solutions and we interpret this partly in the language of
catastrophe theory.

2.   THE BOUSSINESQ EQUATIONS

The two-dimensional steady Boussinesq equations for
buoyancy-driven flow in the untilted cavity are, in non-
dimensional form

$$u \frac{\partial u}{\partial x} + v \frac{\partial u}{\partial y} + \frac{\partial p}{\partial x} - Pr \nabla^2 u = 0$$

$$u \frac{\partial v}{\partial x} + v \frac{\partial v}{\partial y} + \frac{\partial p}{\partial y} - Pr \nabla^2 v - RaPrT = 0$$

$$\frac{\partial u}{\partial x} + \frac{\partial v}{\partial y} = 0$$

$$u \frac{\partial T}{\partial x} + v \frac{\partial T}{\partial y} - \nabla^2 T = 0 \qquad (1)$$

for

$$0 \leqslant x \leqslant \beta, \ 0 \leqslant y \leqslant 1,$$

where u and v are velocity components in the x and y directions, p is pressure, T is temperature, β is the width-to-height ratio of the cavity, Pr is the Prandtl number, and Ra is the Rayleigh number. (In terms of the physical quantities

$$Pr = \frac{\nu}{\kappa}$$

and

$$Ra = \frac{\alpha(\Delta T) \ g \ H^3}{\kappa \nu}$$

where ν is the viscosity of the fluid, κ is the thermal diffusion coefficient of the fluid, α is the coefficient of thermal expansion of the fluid, ΔT is the temperature difference across the cavity, g is the acceleration due to gravity, and H is the height of the cavity). In the case of a tilted cavity the equations have a component of the buoyancy term RaPrT in each of the Navier-Stokes equations of motion. We present here results only for the case when all the cavity walls are rigid. Other boundary conditions have been tackled in the same way [4]. On rigid walls both components of velocity were taken to be zero, while on free walls the normal component only was taken to be zero. The boundary conditions on temperature are

$$T = 0, \quad y = 1,$$

$$T = 1, \quad y = 0,$$

$$\frac{\partial T}{\partial x} = 0, \quad x = 0 \text{ and } x = \beta.$$

Another form of the equations can be obtained by changing variables from x to x* where

$$x^* = x/\beta.$$

The domain of x*, y is then the unit square and is independent of β, which now appears in the equations instead. For example, the first equation of (1) becomes

$$\beta u \frac{\partial u}{\partial x*} + v \frac{\partial u}{\partial y} + \beta \frac{\partial p}{\partial x*} - Pr (\beta^2 \frac{\partial^2 u}{\partial x*^2} \frac{\partial^2 u}{\partial y^2}) = 0 \qquad (2)$$

This formulation has advantages for parameter stepping since the dependence upon the parameter β has been made explicit.

The no-flow state

$$u = 0,$$

$$v = 0,$$

$$p = RaPr(y - \tfrac{1}{2}y^2) + constant,$$

$$T = 1-y \qquad (3)$$

is always a solution of the equations (1) and the boundary conditions.

It may not be the only one and may not always be stable. For high enough Ra other stable solutions appear.

Note that for the untilted cavity if {u(x), v(x), p(x), T(x)} is a solution of equations (1) and the boundary conditions then so is {- u(β-x), v(β-x), p(β-x), T(β-x)}. Thus unless the extra solutions are symmetric about x = β/2 (the centre line) they must occur in pairs.

3.   THE NUMERICAL METHODS

The finite-element method for discretizing equations such as (1) or (2) is standard, and well described elsewhere [5,6,7].

It leads to a set of coupled non-linear algebraic equations for the nodal values of the fields, which can be written

$$F_i (\underline{X}; \underline{a}) = 0, \qquad 1 \leqslant i \leqslant N, \qquad (4)$$

where $\underline{X}$ is the vector of unknown nodal field values (of which there are N), and $\underline{a}$ is the vector of parameters (including Ra, Pr, β and the angle of tilt). In order to solve these equations for given parameter values it is necessary to linearize them and iterate. We use the Newton-Raphson method of linearization about the latest estimate, that is choose an initial guess $\underline{X}^0$ and solve successively for $\underline{X}^1$, $\underline{X}^2$, ... from the linear system

$$\frac{\partial F_i(X^n;\ \underline{a})}{\partial x_j}\ (X_j^{n+1} - X_j^n) = -\ F_i(X^n;\ \underline{a})\ ,\qquad (5)$$

(where we have used the Einstein summation convention). This introduces the jacobian matrix $\underline{\underline{J}}$ where

$$J_{ij} \equiv \frac{\partial F_i}{\partial x_j}\ .$$

We solve the linear system (5) by the (direct) frontal method of Gaussian elimination. $\underline{\underline{J}}$ is decomposed into the product of lower and upper triangular matrices $\underline{\underline{L}}$ and $\underline{\underline{U}}$ as

$$\underline{\underline{J}} = \underline{\underline{L}}\ \underline{\underline{U}}\ ,\qquad (6)$$

and then the equations are easily solved by the usual forward elimination and back substitution processes with triangular matrices. These steps are computationally cheap so the cost of the elimination is dominated by the contribution from the LU decomposition.

The scheme described above has an unexpected bonus. Consider solving (4) for several sets of values $\underline{a}_{(1)}$, $\underline{a}_{(2)}$ ... of the parameters $\underline{a}$. The cost of finding the solution for given values $\underline{a}_{(2)}$ of the parameters is much reduced if the initial guess is good, and an obvious choice is the solution $X_{(1)}$ at neighbouring values $\underline{a}_{(1)}$ of the parameters. An even better guess is

$$X_{(1)i} + \frac{\partial X_{(1)i}}{\partial a_{(1)\alpha}}\ (a_{(2)\alpha} - a_{(1)\alpha})\ ,\qquad (7)$$

and it is not difficult to construct higher interpolations.

Now from (4)

$$\frac{\partial F_i}{\partial x_j}\frac{\partial X_j}{\partial a_\alpha} + \frac{\partial F_i}{\partial a_\alpha} = 0\ .\qquad (8)$$

Thus $\frac{\partial X_j}{\partial a_\alpha}$ is determined by a linear system with the same matrix $\underline{\underline{J}}$ as in (5), but a different right-hand-side, and so once the solution X for given values $\underline{a}$ of the parameter has been determined $\frac{\partial X_j}{\partial a_\alpha}$ can be found very cheaply since it is not necessary to repeat the LU decomposition of $\underline{\underline{J}}$.

144

It is thus possible to obtain a very good initial guess
for the solution at parameter values $a_{(2)}$, once the solution
at values $a_{(1)}$ is known. In many non-linear problems it may
not be possible at some values $a_{(n)}$ of the parameters to find,
a priori, a good-enough initial guess for the Newton-Raphson
iterations to converge. However if it is possible to calculate the solution at one set of values $a_{(o)}$, then by stepping
through the solutions at a series of values between $a_{(o)}$ and
$a_{(n)}$, a good enough initial guess can be obtained. (This is
Davidenko's [8] method of continuation in a parameter). The
parameter stepping can be implemented in many ways and these
schemes make it possible to study very cheaply the behaviour
of the solution to (4) as a function of the parameters. At
first sight there might seem to be a problem if there are any
bifurcations in the solution, but in practice the initial
guess obtained by parameter-stepping is so good that there is
usually no problem in following one branch, just stepping
past bifurcations as illustrated schematically in Figure 2.

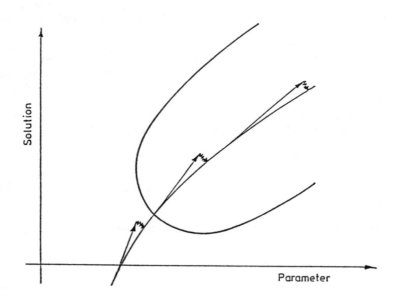

Fig. 2   Schematic diagram of parameter stepping, showing
           its ability to step past bifurcations. The lines
           —— indicate the predictor steps using (7), and
           the lines ∧∧∧〉 indicate the Newton-Raphson steps.

Further, and most importantly, it is actually possible to locate the position of any bifurcation on the branch by monitoring the determinant of the jacobian $\underline{\underline{J}}$. At bifurcations (and limit points) $\underline{\underline{J}}$ becomes rank deficient, which gives rise to a zero of det $|\underline{\underline{J}}|$. Provided that the zero is simple it is easy to locate bifurcations by observing the sign of det $|\underline{\underline{J}}|$, which differs on opposite sides of the bifurcation. Now det $|\underline{\underline{J}}|$ is very easily evaluated because from (6)

$$\det |\underline{\underline{J}}| = (\det |\underline{\underline{L}}|)(\det |\underline{\underline{U}}|) \quad , \tag{9}$$

and since $\underline{\underline{L}}$ and $\underline{\underline{U}}$ are triangular matrices their determinants are just the products of their diagonal entries. In practice the numbers involved are so large that we calculate the sign of det $|\underline{\underline{J}}|$, and the logarithm of the absolute value of det $|\underline{\underline{J}}|$.

Using the above technique it is possible to follow the behaviour of the solution as a parameter changes, and to locate any bifurcations.

It can be shown [9] that an eigenvalue with negative real part of $\underline{\underline{J}}$ for the steady state equations corresponds to instability of the solution as a solution of the time-dependent equations. Since det $|\underline{\underline{J}}|$ is just the product of the eigenvalues of $\underline{\underline{J}}$, then it is easy to see that there is at least one negative eigenvalue of J if det $|\underline{\underline{J}}|$ is negative, and so the solution is unstable. If det $|\underline{\underline{J}}|$ is positive then it is not possible to say that there are no eigenvalues of $\underline{\underline{J}}$ with negative real part, (as there may be an even number of such eigenvalues of course).

In the case of the tilted cavity other techniques have been used [10]. In this case the equations (4) are

$$F_i \ (\underline{X}, \ Ra \ \theta) = 0 \tag{10}$$

with Ra the Rayleigh number and θ the angle of tilt. For $\theta = 0$ we get, as seen later, the usual Bénard pitchfork bifurcation whereas for $\theta \neq 0$ this unfolds as in Figure 3 to a one sided bifurcation (limit point). For the first of these, the bifurcation was found by a method due to Moore [11] which involves solving the equations

$$F(\underline{x}, \ Ra, \ 0) + \Delta\psi = 0 \quad ,$$

$$\psi^T F_x(\underline{x}, \ Ra, \ 0) = 0 \quad ,$$

$$\psi^T F_{Ra}(\underline{x}, \ Ra, \ 0) = 0 \quad ,$$

$$\psi^T \psi = 1 \quad . \tag{11}$$

146

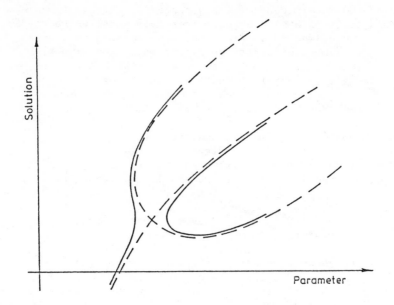

Fig. 3   Schematic diagram showing the effect of the
          perturbation (10) on the bifurcation. The unper-
          turbed bifurcation is shown dashed.

In these equations ψ is the left eigenvector of F_x, a subscript denotes differentiation with respect to that subscript,
and Δ is a parameter introduced for numerical convenience,
since otherwise the system is overdetermined. These equations
were also discretised in the finite-element approximation and
solved by Newton's method. The solution of these equations
was used as the initial guess for the calculation of the
limit point at the first non-zero value of tilt considered.

      The limit (one-sided bifurcation) points were found from
the solution of the extended set of equations proposed by
Moore and Spence [12]

$$F(\underline{x}, \text{ Ra, } \theta) = 0$$

$$F_x(\underline{x}, \text{ Ra, } \theta) \, \phi = 0$$

$$\ell(\phi) = 1 \tag{12}$$

where ϕ is the right eigenvector of F_x and ℓ is a linear
functional which is introduced to normalise ϕ. For a given
value of one parameter, θ say, we again discretise the

equations and solve them by Newton's method to give the value
of Ra at which there is a limit point. We then use Euler-
Newton continuation in the first parameter to obtain the
variation in limit point with θ.

4.   RESULTS: THE SQUARE CAVITY UNTILTED

The above techniques were used to study the bifurcations
from the no-flow solution in the Bénard problem. We illustrate the method of bifurcation search by monitoring the
determinant of the jacobian matrix. We present in Table 1
values of the sign, absolute value and logarithm of the abso-
lute value of det $\underline{\underline{J}}$ at various Ra, for the no-flow solution
to the Bénard problem with rigid walls on all sides, and
width-to-height ratio 1, on a 9x9 nodal grid. It is easy to
see that there are bifurcations between Ra=2650 and Ra=2700

Ra	sign det$\|\underline{\underline{J}}\|$	log(det$\|\underline{\underline{J}}\|$)	det$\|\underline{\underline{J}}\|$
10	+	72.2	2.19×10^{31}
100	+	72.1	2.04×10^{31}
2400	+	68.7	6.89×10^{29}
2500	+	68.2	3.94×10^{29}
2650	+	63.1	2.44×10^{27}
2700	−	66.9	-1.15×10^{29}
3000	−	68.7	-6.97×10^{29}
5000	−	69.3	-7.29×10^{28}
8000	+	68.2	4.29×10^{29}
10000	+	69.0	9.11×10^{29}

Table 1   det$\|\underline{\underline{J}}\|$ versus Ra for the rigid/rigid case, width-
to-height ratio 1, on a 9x9 grid

and between Ra=7000 and Ra=8000. (The negative sign of
det$|\underline{\underline{J}}|$ shows that the no-flow state must be unstable between
Ra=2700 and Ra=7000). Values of det$|\underline{\underline{J}}|$ versus Ra for values
of Ra between 2400 and 2700 are shown in the graph of Figure
4. It is possible to locate the position of the bifurcation
by interpolation between values of det$|\underline{\underline{J}}|$ on either side of
the bifurcation, leading to the value 2652. This corresponds
to Ra$_{crit}$ of course.

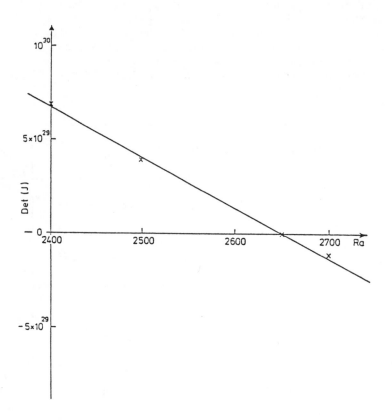

Fig. 4   Det $\underline{\underline{J}}$ versus Ra for the rigid/rigid case, for
          width-to-height ratio 1, on a 9x9 grid, for Ra in
          the range 2400 to 2700. The straight line is a
          fit by eye to the values.

The calculation was repeated on a 17x17 grid in order
to check the dependence of the bifurcation value on the grid
spacing. The first bifurcation on a 17x17 grid was at
Ra=2612. Thus a quite coarse mesh produced very accurate
results.

A perturbation technique [4] was used to find the other branches. At Ra=2650 they were thereby identified as one-cell solutions, and the other branches leading from the bifurcation at Ra=7128 were identified as two-cell solutions.

Figure 5 shows streamlines for the one-cell solution at Ra=10,000, on a 17x17 grid and Figure 6 shows the isotherms. Figure 7 shows the streamlines for the two-cell solution at Ra=10,000 on a 17x17 grid and Figure 8 shows the isotherms.

5.   RESULTS: THE TILTED CAVITY

The pitchfork bifurcation in one parameter, Ra, in the usual case is unstable in the sense that small perturbations to the boundary conditions or equations, which break the symmetry, change the qualitative picture. By introducing a second, symmetry-breaking parameter into the problem, the pitchfork is unfolded to a cusp catastrophe which is structurally stable [13]. The original pitchfork is just a section through the cusp.

From these considerations, we deduce that, when the surfaces in the Bénard problem are tilted at θ degrees to the horizontal, then the pitchfork is unfolded to a cusp catastrophe in the parameters Ra and θ. This is shown in Figure 9, from which it is clear that the section at θ=0 through the cusp is the usual Bénard bifurcation diagram. For non-zero θ, a typical section gives the one-sided bifurcation shown in Figure 10. In this case the flow develops smoothly from zero Rayleigh number to the lower branch. There are however two further solutions above a critical value of Ra which cannot be reached smoothly. The lower of these is unstable, but the upper branch is stable and is in principal observable.

Figure 10 shows a state diagram generated by the techniques described above, for a 1° angle of tilt. The circles are computed values and have been joined smoothly by a spline interpolation. The labels ±1 are the predicted values of the Leray-Schauder index [14] and these are in agreement with the expected stability properties, a negative sign indicating unstable steady flow. In the present case, the branches with positive index are stable. The anomalous solution on the upper branch is similar to the normal solution but with opposite sense of rotation. The anomalous streamlines and isotherms at a Rayleigh number of 5000 are shown in Figure 11.

Figure 12 shows the locus of limit points as a function of Ra and θ. This is the projection of the folds of Figure 1 onto the Ra-θ plane and shows the critical Rayleigh number for the appearance of the anomalous solution at any tilt. It is of course symmetric about θ=0, and the cusp is evident

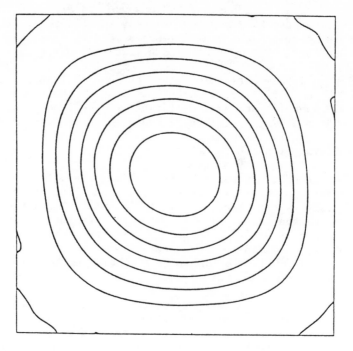

Fig. 5   Streamlines for the one-cell solution on Ra=10,000
           on a 17x17 grid for the rigid/rigid case, at width-
           to-height ratio 1.

Fig. 6   Isotherms for the one-cell solution of Figure 8.

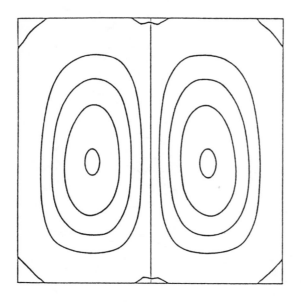

Fig. 7   Streamlines at Ra=10,000 for the two-cell solution
          on a 17x17 grid for the rigid/rigid case, at width-
          to-height ratio 1.

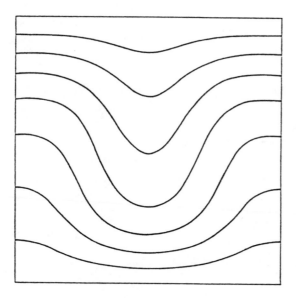

Fig. 8   Isotherms for the two-cell solution at Ra=10,000,
          on a 17x17 grid for the rigid/rigid case, at width-
          to-height ratio 1.

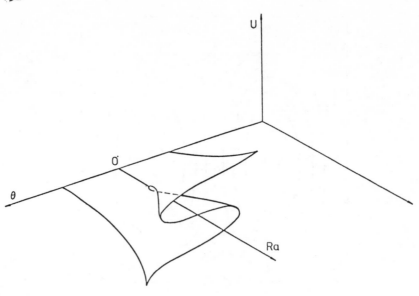

Fig. 9 The cusp catastrophe for Bénard convection in a
tilted cavity.

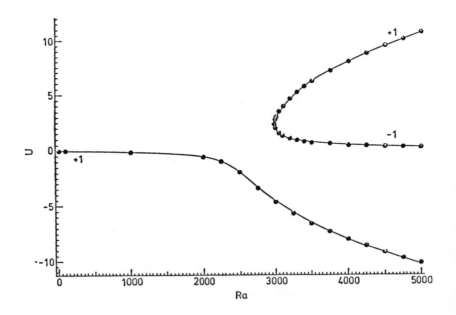

Fig. 10 State diagram for cavity tilted by 1 degree. The
component of velocity parallel to the top of the
cavity at (0.5, 0.9) is plotted against the Rayleigh number.

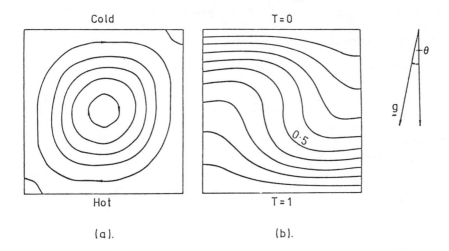

Fig. 11   Streamlines (a) and isotherms (b) for the anomalous
solution at a Rayleigh number of 5000 and 1 degree
tilt.

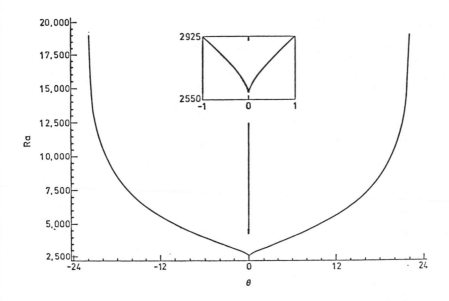

Fig. 12   Locus of limit points as a function of Ra and θ.

154

over a rather narrow range of angles, of $\pm 1^{\circ}$. However, the striking feature of Figure 12 is that the curve is asymptotic to the lines $\theta = \pm 22.0^{\circ}$, so that the anomalous solutions exist only for angles of tilt less than this limiting value.

6. RESULTS: THE VARIATION WITH ASPECT RATIO

A detailed study was made [4] of the Bénard problem with rigid side walls for width-to-height ratios in the range 1 to 4. The bifurcations from the no-flow branch were determined as above, by monitoring $\det|\underline{\underline{J}}|$ on this branch. The results are presented in Table 2 and Figure 13. Results of Luijkx

Width-to-height Ratio	Grid	Bifurcation values of Ra			
1.0	5x5	3091	9812	16139	41223
"	9x9	2665	7130	21406	25032
"	17x17	2588	6774		
"	21x21	2612	6491		
1.2	9x9	2353	4432	11684	
1.4	9x9	2332	3190	7286	
1.6	9x9	2438	2564	5039	
"	17x9	2434	2471	5027	
"	33x9	2433	2462		
1.8	9x9	2227	2588	3829	
"	17x9	2218	2591	3917	
"	33x9	2202	2583		
1.9	17x9	2122	2653	3607	
2.0	9x9	2049	2563	3372	
"	17x9	2056	2650	3461	
"	33x9		2629	3427	
2.2	9x9	1961	2250	3534	
"	17x9	1988	2453	3573	
2.4	9x9	1947	1956	3882	
"	17x9	1976	2236	3509	3882
2.6	9x9	1740	1942		
"	17x9	2000	2079	3022	4095
2.8	9x9	1576	1981		
"	17x9	1978	2038	2696	
3.0	9x9	1450			
"	17x9	1916	2071	2504	
"	33x9	1901	2060		
3.2	33x9	1872	2038		
3.4	33x9	1866	1977		
3.6	33x9	1875	1917		
3.8	33x9	1874	1891		
4.0	33x9	1844	1908		

Table 2   Lowest few bifurcation values of Ra for the rigid/ rigid case

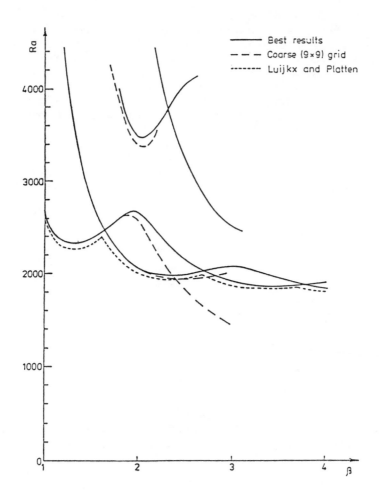

Fig. 13   Bifurcation values versus width-to-height ratio.
The insert shows the actual behaviour in the
encircled region.

and Platten [15] for the rigid/rigid case are also presented
for comparison.

As the width-to-height ratio is increased the lowest
bifurcation value of Ra changes from a branch corresponding
to a one-cell solution to a branch corresponding to a two-
cell solution, and then to a branch corresponding to a three-
cell solution, and so on. The separation between the minima
of the various branches is nearly constant for each boundary
condition. This implies that there is a preferred convection
cell width of about 1.02. The analytic values for an infinite width-to-height ratio is also 1.02.

156

One feature of our results deserves comment. We
expected that the bifurcation diagrams should be as shown in
Figure 14. However the curves in the encircled area do not
cross in practice but are as shown in the insert and in
Figure 13. This is because the "one-cell" solution is actually
a mixture of one and three cell solutions. One of the three
cells dominates at low values of β but two others appear and
grow as β is increased until when β ⊵ 3 this solution looks
like a "three-cell" solution.

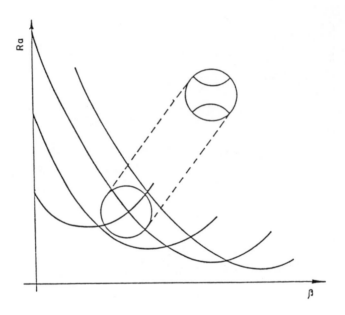

Fig. 14   Expected plot of bifurcations versus width-to-
           height ratio. The insert shows the actual
           behaviour in the encircled region.

The events of Figure 13 can also be interpreted in terms
of catastrophe theory. At each point on a line of Figure 13,
other than where lines cross, there is a pitchfork bifurcation of the usual kind illustrated in Figure 1. The extra
parameter β here does not break the symmetry so the pitchfork is not unfolded to a cusp unless the cavity is tilted
for example. At crossing points things are much more complicated. The Jacobian matrix now has a doubly-denenerate
eigen-value of zero and this leads to a double-cusp catastrophe. Unfolding this needs a total of 8 parameters, rather
than the 2 for the square cavity, and is not yet fully worked
out. Apart from lack of crossing due to mixing the whole
analysis for the Bénard problem is analogous to that done for

the plate buckling problem in reference 13.

ACKNOWLEDGEMENTS

This paper is based largely on recent work of Drs. K.H. Winters, C.P. Jackson and Mr. K.A. Cliffe at Harwell. I thank them for discussions on the material and their agreement to allow me to present it as a lecture to the 3rd International Conference on Numerical Methods in Thermal Problems, Seattle, August 1983.

REFERENCES

1. RAYLEIGH, LORD, On Convective Currents in a Horizontal Layer of Fluid when the Higher Temperature is on the Underside, Phil. Mag. 32, 529-547, 1916.

2. CHANDRASEKAR, S., Hydrodynamic and Hydromagnetic Instability, Oxford Univ. Press, 1961.

3. PELLOW, A. and SOUTHWELL, R.V., On Maintained Convective Motion in a Fluid Heated from below, Proc. Roy. Soc. A., 176, 312-343, 1940.

4. JACKSON, C.P. and WINTERS, K.H., A Finite Element Study of the Bénard Problem using Parameter-stepping and Bifurcation Search, AERE Report AERE-TP.936, 1982.

5. ZIENKIEWICZ, O.C., The Finite-Element Method, McGraw-Hill, London, 1977.

6. JACKSON, C.P. and CLIFFE, K.A., Mixed Interpolation in Primitive Variable Finite-Element Formulations for Incompressible Flow, Int. J. Num. Meth. Eng., 17, 1659-1688, 1981.

7. GRESHO, P.M., LEE, R.L., CHAN, S.T. and LEONE, J.M., A New Finite Element for Incompressible or Boussinesq Fluids in Proceedings of the Third International Conference on Finite Elements in Flow Problems, Banff, Canada, 1980.

8. DAVIDENKO, D.F., On a New Method of Numerically Integrating a System of Nonlinear Equations, Dokl. Acad. Nauk SSSR 88, 601-602, 1953.

9. CLIFFE, K.A. and GREENFIELD, Some Comments on Laminar Flow in Symmetric Two-Dimensional Chennels, AERE Harwell Report, AERE-TP.939, 1983.

10. CLIFFE, K.A. and WINTERS, K.H., A Numerical Study of the Cusp Catastrophe for Bénard Convection in Tilted

Cavities, AERE Harwell Report AERE-TP.997, 1983.

11. MOORE, G., The Numerical Treatment of Non-Trivial Bi-
    furcation Points, Numer. Funct. Anal. and Optimiz. 2,
    441-472, 1980.

12. MOORE, G. and SPENCE, A., The Calculation of Turning
    Points of Non-Linear Equations, SIAM J. Numer. Anal. 17,
    567-576, 1980.

13. POSTON, T. and STEWART, I.N., Catastrophe Theory and
    its Applications, London: Pitman, 1978.

14. BENJAMIN, T.B., Applications of Leray-Schauder Degree
    Theory to Problems of Hydrodynamic Stability, Math.
    Proc. Camb. Phil. Soc. 79, 373-392, 1976.

15. LUIJKX, J. and PLATTEN, J.K., On the Onset of Free Con-
    vection in a Rectangular Channel, J. Non-Equilib.
    Thermodyn. 6, 141-158, 1981.

CHAPTER 7

COMBUSTION AND EXPLOSION EQUATIONS AND THEIR CALCULATION

Karl Gustafson

University of Colorado

INTRODUCTION

Our discussion will be based to some extent on [1,2,3], in which we obtain high precision numerical calculation of the S-shaped bifurcation diagrams and the corresponding positive temperature solutions for a class of equations occurring in combustion and explosion theory. More specifically, the equations dealt with in [1,2,3] are those of spontaneous ignition of reactive solids obeying zero or low order kinetic descriptions. Only spherical geometries were considered in [1,2,3]. The numerical scheme, which we call HOC, was shooting combined with a higher order calculus. This improved previous treatments, and the emphasis was on obtaining extremely sharp methods for the calculation of the values of the critical parameters of ignition.

The prototypical equation to be kept in mind during this discussion is

$$-\Delta u = \lambda e^{\frac{u}{1+\varepsilon u}} \quad , \quad \lambda \geq 0 \quad , \quad \varepsilon \geq 0 \quad , \tag{1}$$

where u is a temperature in a self heating body Ω near explosion, λ represents the inherent exothermicity of the substance, and ε^{-1} the activation energy. There are fundamental qualitative differences between the case $\varepsilon > 0$, which is sometimes called the full Arrhenius model, and the case $\varepsilon = 0$, which is sometimes called the Frank-Kamenetskii model. Although the lower solution branches of both are quite close, for the former a situation prevails whereby the solutions exist for all λ whereas for the latter there is a critical value λ_c beyond which no solutions exist. We will return to this distinction and its effect on the interpretations of ignition and explosion later.

162

Because we know of no presentation which aims to relate and, shall we say, bounce off one another,

(1) Physical Issues
(2) Analytical Issues
(3) Numerical Issues
(4) Historical Issues

concerning such equations, we shall do so in this survey. Our aim is not only to present a partial survey of state of the art numerical methods for combustion problems such as (1) but also to relate those quantitative aspects to certain qualitative ones. In a sense then this presentation may be regarded as partially tutorial. A more extensive treatment will be given in [4].*

CONTENTS

NOTATION

Δ - Laplacian
λ - exothermicity, eigenvalue parameter
u - temperature perturbation
ε - inverse activation energy, β^{-1}
n - spatial dimension
j - $n-1$
r - radial distance
ϕ - linearized eigenfunction
C - concentration
T - temperature
k - diffusion coefficient
q_o - latent heat of reaction
a_o - activation energy constant
$\|u\|$ - norm of u , $u(0)$, A
$G(P,Q)$ - Green's function
Ω - region
$\partial\Omega$ - boundary of region
α - shooting parameter
Le - Lewis number

*
Work partially supported by NSF Grant MCS 80-12220A2.

1. THREE NUMERICAL APPROACHES

For the numerical resolution of equations such as (1) as they occur in combustion and explosion, let us identify three principal numerical approaches that may be, and have been, taken. Our intent in this section is to proceed quickly in an a priori way to the numerical treatments, discussing a few qualitative or quantitative issues as they arise. This section is thus one of preview and formulation. References and details will be postponed to later sections of this survey.

We gave no boundary conditions with (1) but what we will have in mind usually will be Dirichlet boundary conditions. Other boundary conditions such as Neumann and especially Robin (finite Biot number) are also important for applications. The Dirichlet boundary condition for u often means $u = T - T_o$

where T_o is a fixed temperature rather close to a critical temperature T at which ignition, explosion, extinction, or some similar phenomenon occurs. Usually we neglect the process of assembly, namely, the bringing together of the chemical constituents for the reaction.

For nonlinear elliptic boundary value problems such as

$$\begin{cases} -\Delta u = \lambda e^{\frac{u}{1+\varepsilon u}} & \text{in } \Omega \\ u = 0 & \text{on } \partial\Omega \end{cases} \tag{1.1}$$

we may delineate three classes of numerical schemes in use:

(1) Shooting with ODE solvers;
(2) Associated linear eigenvalue solutions;
(3) Arc length continuation schemes.

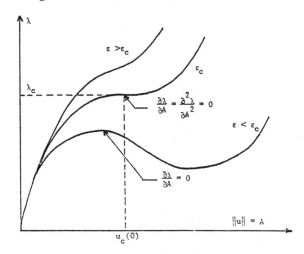

Figure 1. Bifurcation Diagrams and the Critical Point.

All have successfully obtained the bifurcation diagrams depicted in Figure 1.

Approach (1) solves (1.1) for suitable domains by means of ODE Solvers combined with Newton-Raphson techniques. For spherical domains (1.1) becomes

$$
\begin{cases}
-\left(\dfrac{d^2 u}{dr^2} + \dfrac{j}{r}\dfrac{du}{dr}\right) = \lambda e^{\frac{u}{1+\varepsilon u}} \ , \quad 0 < r < 1 \ , \\[4mm]
u'(0) = u(1) = 0 \ .
\end{cases}
\tag{1.2}
$$

For given ε it is best to shoot from an assigned $u(0) = A$, obtaining λ by iteration through a boundary expansion of the condition $u(1) = 0$. This is an old trick in bifurcation theory to avoid the vertical slopes that would result if instead one starts with λ. Approach (1) has thus far apparently been limited to so called class A geometries: in particular, slab (dim = 1, j = 0), cylinder (dim = 2, j = 1), and sphere (dim = 3, j = 2) geometries. It is the approach we will emphasize here. It yields very high precision (roughly: as high as you want) criticality parameters λ_c, u_c, ε_c, and has an advantage of providing at the same time very accurate temperature solution profiles. The critical values that we obtained are given in Table 1.

	λ_c	u_c	ε_c
Slab	1.307374	4.896548	0.245780
Cylinder	3.006301	5.943244	0.242106
Sphere	5.041112	7.184944	0.238797

Table 1.   Critical Arrhenius Parameters.

Approach (2) is based upon finding the critical value λ_c as the first eigenvalue λ_1 of the linearization of (1.1) about u_c, namely

$$
\begin{cases}
-\Delta\phi = \lambda\left((1+\varepsilon u_c)^{-2}\, e^{\frac{u_c}{1+\varepsilon u_c}}\right)\phi \quad \text{in } \Omega \ , \\[4mm]
\phi = 0 \qquad\qquad\qquad\qquad \text{on } \partial\Omega \ .
\end{cases}
\tag{1.3}
$$

Usually in this approach λ_1, u_c, and ϕ are obtained by a pseudo-transient method. This approach gives very good critical parameter values whose quality and efficiency of calculation depend to some extent on that of the time-dependent scheme employed. In practice it has been implemented primarily for the class A geometries.

Approach (3) uses arc length continuation schemes whereby one advances in increments along the bifurcation curve of the particular problem, that is, the curve of the solution size $\|u\|$ plotted as a function of the bifurcation parameter λ . It is well suited to the use of fast elliptic solvers for an intermediate linear calculation such as (in Frank-Kamenetskii approximation)

$$\begin{cases} -\Delta u_{n+1} = \lambda e^{u_n} & \text{in } \Omega \\ u_{n+1} = 0 & \text{on } \partial\Omega \end{cases} \tag{1.4}$$

and thus has an advantage of being applicable in principle to general domains. There have been some difficulties employing it exactly or near criticality.

We do not mean to imply that these are the only numerical approaches that have been employed for problems of this type. All manner of elliptic and iterative methods are fair game, and for the pseudo-unsteady approaches many parabolic and marching schemes have been considered. Several types of finite difference and finite element schemes, some to be mentioned later, have been employed. We are not aware of any calculations by means of spectral methods but that does not rule them out.

2.   INTUITION FOR THE EQUATIONS

Equations such as (1.1) arise in chemical reaction kinetics as systems of the form

$$\begin{cases} (C_i)_t = k_i \Delta C_i + f_i(C_j, T) \\ T_t = k \Delta T + f(C_j, T) \end{cases} \tag{2.1}$$

along with appropriate initial and boundary values. In (2.1) the C_j are concentrations, the k_i are diffusion coefficients for the various agents, the f_i are generally nonlinear reaction terms such as the right hand side of (1.1), and T is the temperature. Complex reactions are treated by decoupling them as much as possible into elementary reaction steps. A central question in reaction kinetics is then the understanding of these elementary reaction steps and the calculation of their critical parameters. Assuming a constant concentration over a short time interval yields the single nonlinear equation

$$T_t = k_o \Delta T + q_o c_o e^{-a_o/T} \tag{2.2}$$

where q_o is the latent heat of reaction, c_o the concentration and a_o the activation energy for the substance.

Assuming that the temperature T does not exceed too much a base temperature T_o already near criticality, we may expand the Arrhenius exponent in the Neumann series

$$\frac{1}{T} = \frac{1}{T_o} \left(1 + \frac{T-T_o}{T_o}\right)^{-1}$$

$$= \frac{1}{T_o} - \frac{(T-T_o)}{T_o^2} + \frac{(T-T_o)^2}{T_o^3} \quad \cdots \qquad (2.3)$$

Letting u denote the perturbation temperature $(T-T_o)a_o/T_o^2$, assuming that the surface $\partial\Omega$ of the body Ω remains temporarily at constant temperature T_o, and dropping all but the linear terms in the exponent (2.3), we have

$$u_t = k_o \Delta u + a_o T_o^{-2} q_o c_o e^{-a_o/T_o} e^u . \qquad (2.4)$$

For calculating critical parameters we will be particularly interested in the steady equilibrium heat power balance equation when $u_t = 0$, which we may write from (2.4) as

$$-\Delta u = \lambda e^u \qquad (2.5)$$

where $\lambda = k_o^{-1} a_o T_o^{-2} q_o c_o e^{-a_o/T_o}$.

Equations (2.4) and (2.5) are the Frank-Kamenetskii approximation, which corresponds as seen above to cutting off the Neumann series after the first two terms. For small u this permits a considerable simplification in the nonlinearity in the problem. On the other hand, by means of the identity

$$-\frac{a_o}{T} = \frac{u}{1+\varepsilon u} - \frac{a_o}{T_o} \qquad (2.6)$$

where $\varepsilon = T_o/a_o$, the full equation (2.2), which we shall refer to as the (full) Arrhenius equation, becomes

$$u_t = k_o \Delta u + a_o T_o^{-2} q_o c_o e^{-a_o/T_o} e^{\frac{u}{1+\varepsilon u}} \qquad (2.7)$$

and in particular the Arrhenius equation in the steady case is

$$-\Delta u = \lambda e^{\frac{u}{1+\varepsilon u}} \qquad (2.8)$$

as given in (1.1).

Note that, given the other assumptions leading to these models, the untruncated Arrhenius equations (2.7) and (2.8) are valid for all $|T-T_o| < |T_o|$, i.e., for perturbation temperatures $0 < u < \varepsilon^{-1}$, whereas the Frank-Kamenetskii equations (2.4) and (2.5) steadily lose precision as the exothermic reaction raises the temperature above T_o . Roughly, the Frank-Kamenetskii equation describes a chemical with an infinitely large activation energy when assembled at the temperature T_o .

The role of equation (2.8) in ignition and explosion theory may be easily visualized in terms of the activation parameter ε . As heat is generated in the body, the heat diffused within the body via the left side (conduction) of equation (2.8) must be sufficient to balance that generated by the right side (reaction) of equation (2.8). Holding the boundary temperature (sometimes called the supply temperature) constant at T_o , in the low temperature limit $u \to 0$ we have

$$\lambda e^{\frac{u}{1+\varepsilon u}} \sim \frac{a_o}{T_o^2} \cdot \frac{q_o c_o}{k_o} \cdot e^{-a_o/T_o} \tag{2.9}$$

and in the high temperature limit $u \to \infty$

$$\lambda e^{\frac{u}{1+\varepsilon u}} \sim \frac{a_o}{T_o^2} \cdot \frac{q_o c_o}{k_o} . \tag{2.10}$$

Letting (r.h.s.) denote either of the right sides of (2.9) or (2.10), the boundary value problem

$$\begin{cases} -u_{xx} = (r.h.s.) & -1 < x < 1 \\ u(-1) = u(1) = 0 \end{cases} \tag{2.11}$$

with solution

$$u(x) = \frac{(r.h.s.)}{2} (1-x^2)^2 \qquad -1 \le x \le 1 \tag{2.12}$$

models the exothermic temperature buildup. A smaller diffusion coefficient k_o relative to the activation energy a_o produces more internal heating.

To continue the heuristics, note that from the identity (2.6) one has

$$(r.h.s.: u \to 0) < \lambda e^{\frac{u}{1+\varepsilon u}} < (r.h.s.: u \to \infty) \tag{2.13}$$

and thus we expect the true solution of the nonlinear equation (2.8) to be bounded below by the linear low temperature approximation and above by the linear high temperature approximation. Similarly from the relation

$$\lambda e^{\frac{u}{1+\epsilon maxu}} < \lambda e^{\frac{u}{1+\epsilon u}} < \lambda e^{\frac{u}{1+\epsilon minu}} \qquad (2.14)$$

we expect the true solution to be bounded below and above by the solutions of the nonlinear equations corresponding to the bounds of (2.14), the upper being exactly the Frank-Kamenetskii approximation.

In theory one can construct all solutions to equations of the form

$$\begin{cases} -\Delta u = f(u,\lambda,\epsilon) & \text{in} \quad \Omega \\ u = 0 & \text{on} \quad \partial\Omega \end{cases} \qquad (2.15)$$

by means of the itegral equation

$$u(P) = \int_{\Omega} G(P,Q)f(u(Q),\lambda,\epsilon)d\Omega . \qquad (2.16)$$

Because, given some differentiability of the nonlinearity f, (2.15) and (2.16) are equivalent the solution of (2.15) and (2.16) will prevail for any general boundary condition and any domain Ω in any number of dimensions provided that the Green's function $G(P,Q)$ for that domain and those boundary conditions exists. As is well known, this is rather generally assumed. Moreover, then one should be able to iterate (2.16) to a solution, namely, by

$$u^{n+1}(P) = \int_{\Omega} G(P,Q)f(u^n(Q),\lambda,\epsilon)d\Omega . \qquad (2.17)$$

Let us illustrate, following [5], this rather straightforward and elementary approach. For simplicity we consider the Frank-Kamenetskii case (2.5). The first practical difficulty is to enumerate the Green's function $G(P,Q)$, whose existence far exceeds its accessibility. For the interval $\Omega = (-1,1)$ we have

$$G(x,s) = \begin{cases} \frac{1}{2} (1+s)(1-x) , & s \leq x \\ \frac{1}{2} (1-s)(1+x) , & s \geq x \end{cases} \qquad (2.18)$$

and as first approximation we take the linear interpolate of the Dirichlet boundary condition, namely

$$u^o(x) \equiv 0 , \qquad -1 < x < 1 \qquad (2.19)$$

From this, via (2.17),

$$u^1(x) = \frac{\lambda}{2}(1-x^2)$$

(2.20)

$$u^2(x) = 1 - e^{\frac{\lambda}{2}(1-x^2)} + \frac{\lambda}{2}e^{\frac{\lambda}{2}}\left[\int_{-1}^{1}e^{1-\frac{\lambda}{2}s^2}ds\right.$$

$$\left. + x\left(-\int_{-1}^{x}e^{-\frac{\lambda}{2}s^2}ds + \int_{1}^{x}e^{-\frac{\lambda}{2}s^2}ds\right)\right]$$

and so on.

Given the positivity of the nonlinearity f, which we usually will tacitly assume, from (2.16) one sees immediately that the positivity of solutions u (i.e., exothermicity of the reaction) follows immediately from the positivity of the Green's function. Should the above iterates converge, one will obtain a solution. From the symmetry of the iterates, the solution will be symmetric about the center temperatures.

3. QUALITATIVE FEATURES

It can be shown rigorously, see Amann [6,7], under general conditions for a wide class of elliptic nonlinear eigenvalue problems that if one exhibits both an upper solution w_u, and a lower solution w_ℓ, namely, functions w_u and w_ℓ satisfying

$$w_u(P) \geq \int_\Omega G(P,Q)f(w_u(Q),\lambda,\varepsilon)d\Omega$$

$$w_\ell(P) \leq \int_\Omega G(P,Q)f(w_\ell(Q),\lambda,\varepsilon)d\Omega$$

(3.1)

then a solution u to (2.15) will exist and either w_u or w_ℓ may be taken as initial iteration u_o which will converge to u under the method of successive approximations (2.17). Clearly $w_\ell \equiv 0$ is a lower solution in the example given above. Letting $w_u = w/\max|w|$ where w is the solution of $-\Delta w = 1$ in Ω, $w = 0$ on $\partial\Omega$, produces an upper solution.

Let us show that solutions are symmetric in the case of one dimension. Here we are following Laetsch [8]. We consider

$$\begin{cases} -u''(x) = \lambda f(u), & -1 < x < 1, \\ u(-1) = u(1) = 0, \end{cases}$$

(3.2)

with $\lambda > 0$, $f(v)$ strictly positive, monotone increasing, and continuous.

1. It is easily verified that u is positive, concave, with unique maximum. Call this maximum any of the notations $\|u\| \equiv u(x_o) = \max\{u(x): -1 \le x \le 1\}$. For $-1 \le x < x_o$, $u'(x) > 0$; for $x_o < x \le 1$, $u'(x) < 0$.

2. Integrating $-u'(x)u''(x) = \lambda u' f(u)$, we have for any y in the interval $(-1,1)$

$$- \frac{1}{2} (u'(s))^2 \Big|_{x_o}^{y} = \lambda F(u(s)) \Big|_{x_o}^{y} \tag{3.3}$$

where F is any antiderivative of f . Thus

$$\frac{u'(y)}{\sqrt{F(\|u\|) - F(u(y))}} = \pm (2\lambda)^{1/2}. \tag{3.4}$$

3. We now proceed by cases.

a) For $-1 \le y < x_o$, take the (+) case in (3.4) and with change of variable $v = u(y)$, $dv = u'(y)dy$ upon integration of (3.4) we obtain

$$\int_0^{u(x)} \frac{dv}{\sqrt{F(\|u\| - F(v)}} = (2\lambda)^{1/2}(1+x) . \tag{3.5}$$

b) Similarly from $x_o < y < 1$ we have

$$\int_0^{u(x)} \frac{dv}{\sqrt{F(\|u\|)-F(v)}} = (2\lambda)^{1/2}(1-x) . \tag{3.6}$$

4. From a) and b) at $x = x_o$ we have

$$(2\lambda)^{1/2}(1+x_o) = (2\lambda)^{1/2}(1-x_o) . \tag{3.7}$$

Thus $x_o = 0$. It follows that u is symmetric.

It was not until 1979 that Gidas, Ni, and Nirenberg [9] showed that in three and higher dimensions these conditions guarantee that all solutions are symmetric if one is on a symmetric domain. The key element utilized in [9] is the maximum principal for elliptic equations.

Few who are uninitiated would guess that

$$\begin{cases} -\Delta u = \lambda e^u , & r < 1 \\ u(r) = 0 & \text{at} \quad r = 1 \end{cases} \tag{3.8}$$

has two solutions for $0 < \lambda < \lambda_c$ and no solutions for $\lambda > \lambda_c$,

in the case of one and two dimensions, and that for the physically most important case of three dimensions the situation is

$$0 < \lambda_{1c} <...< \lambda_{(2n+1)c} <...\lambda_\infty <...< \lambda_{(2n)c} <...< \lambda_{2c} < \lambda_c \quad (3.9)$$

wherein there is one solution, three solutions, five solutions,
and so on to the left of the asymptote λ_∞ , where there is an

infinite number of solutions, the number of solutions then
decreasing in even numbers of solutions to two and then none.
We have depicted the situation in Figure 2. We will show later
how this qualitative behavior may be proved, and then at the
end how all of the critical points may be calculated. Such
calculations are at current state of the art research and the
returns are not yet all in.

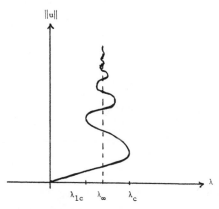

Figure 2. Screw-Shaped Bifurcation Diagram.
The true Arrhenius equation

$$\begin{cases} -\Delta u = \lambda e^{\frac{u}{1+\varepsilon u}}, & r < 1 \\[2mm] u(r) = 0 & \text{at} \quad r = 1 \end{cases} \quad (3.10)$$

has bifurcation curve similar to that of the Frank-Kamenetskii
approximation (3.8) depicted in Figure 2 but as mentioned
earlier qualitatively differs in a fundamental way because
solutions u exist for all positive λ . This is easily
verified by the critieria of Keller and Cohen [10]. Thus for
each $\varepsilon > 0$ the bifurcation curve for (3.10) must eventually
and irrevocably turn to the right. As ε increases the
"wiggles" are damped and at a critical value ε_c one achieves
the situation depicted in Figure 3 wherein the branching
ceases. That is, for each $\varepsilon > \varepsilon_c$ there is only one branch
u(λ) of solutions as parametrized by the bifurcation parameter
λ . Exact values of ε_c, λ_c, and u_c were given in Section 1
for the slab, cylinder, and sphere geometries. Others will be
mentioned later.

172

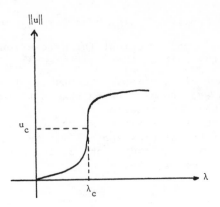

Figure 3. Exact Criticality.

Here is a way to restore intuition in quickly seeing the existence of multiple solutions. We will cheat and use an exact form of the solution known in the one-dimensional case. See Gelfand [11] for a more general treatment. For

$$
\begin{cases}
-u'' = \lambda e^{u} & -1 < x < 1 \\
u(-1) = u(1) = 0
\end{cases}
\tag{3.11}
$$

let $e^{v} = 1/\cosh^{2}x$, so that $v(x) = -2\ell n \cosh x$. Let $u(x) = \beta + v(e^{\alpha/2}x) = \beta - 2\ell n \cosh(e^{\alpha/2}x)$. Note that u is symmetric with two constants α and β which may be used to attempt satisfaction of the boundary conditions. We proceed as follows.

1. Note that

$$
\lambda e^{u(x)} = \lambda e^{\beta}(\cosh(e^{\alpha/2}x))^{-2}
$$

and that

$$
u'' = -((-2e^{\alpha/2}\cosh(e^{\alpha/2}x))^{-1}\sinh(e^{\alpha/2}x)' \tag{3.12}
$$

$$
= 2e^{\alpha/2}(\cosh(e^{\alpha/2}x))^{-2}.
$$

Hence u is a two parameter family of solutions of the differential equation, constrained by λ according to $\lambda = 2e^{\alpha-\beta}$

2. From the boundary condition $u(1) = 0$ and **1)** we have

$$\alpha + \ln(2/\lambda) = \beta = \ln(\cosh e^{\alpha/2})^2 \qquad (3.13)$$

and hence $e^{\alpha} = (\lambda/2)(\cosh e^{\alpha/2})^2$. Letting $e^{\alpha/2} = r$,

$$r = (\lambda/2)^{1/2}\cosh r . \qquad (3.14)$$

This provides us with the root diagram shown in Figure 4.

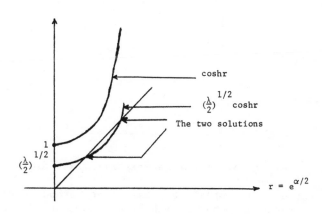

Figure 4.   Root Diagram.

Notice how this shows the restriction on λ : one must have λ small so as to keep the $(\lambda/2)^{1/2}\cosh r$ curve down to where it may intersect the $45°$ line. For the intersection points, let $r = \ell nz$ and solve for the positive zeros of

$$z^2 - (2^{3/2}/\lambda^{1/2})z\ell nz + 1 = 0 \qquad (3.15)$$

from which

$$\alpha_\ell = 2 \ln \ln z_\ell \implies \text{upper branch soln}$$

$$\alpha_u = 2 \ln \ln z_u \implies \text{lower branch soln} \qquad (3.16)$$

$$\beta_{\ell,u} = \ln(\cosh \ln z_{\ell,u})^2$$

3. The dependence of the two solutions on λ may be worked out from the above relations. In particular, note that

$$\lambda \to 0 \implies r_\ell \to 0 \implies \beta_\ell \to 0$$
$$\lambda \to 0 \implies r_u \to \infty \implies \beta_u \to \infty \qquad (3.17)$$

which shows that the bifurcation diagram continues asymptotically upward without limit on the upper branch.

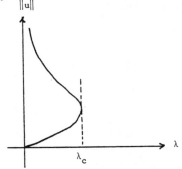

Figure 5.   Frank-Kamenetskii Case, dim n = 1 .

Let us next apply the phase plane method to demonstrate the screw-shaped behavior exhibited in Figure 2. This approach as far as we know goes far back to Emden's theory of stars [12]. In particular this proves for certain isothermal gas stars that there are an infinite number of equilibria positions. For our combustion context it shows that there are an infinite number of exothermic solutions at the parameter value λ_∞ in three dimensions. For simplicity we consider only the Frank-Kamenetskii equation (3.8). The full Arrhenius equation (3.10) may be treated similarly. Exact bifurcation curves calculated numerically for both (3.8) and (3.10) may be found in [1,4], among others.

A point we would like to make here is that the analytic phase plane method and the use of ODE Solvers for numerical solution of these problems are conceptually the same. The ODE Solver reduces the second order equation to a system of first order equations and then follows their trajectories. The higher order calculus scheme (HOC) we introduced in [1] may be viewed as a refinement to higher order of the phase plane trajectories.

Turning then to (3.8) and by use of the now known fact that all solutions of this partial differential equation are spherically symmetric, we may write it as the ordinary differential equation

$$
\begin{cases}
\dfrac{1}{r^2} \dfrac{d}{dr} \left(r^2 \dfrac{du}{dr} \right) + \lambda e^u = 0 , & 0 < r < 1 \\[2em]
u'(0) = u(1) = 0 .
\end{cases}
\tag{3.18}
$$

The fundamental singularity $u_\infty(r) = -2\ln r$ satisfies the differential equation of (3.18) at $\lambda_\infty = 2$ and also the boundary condition $u_\infty(1) = 0$. It may be used to set up a phase plane analysis in which it plays the role of the time variable. Note that as r goes from 1 to 0 , the $u_\infty(r)$ values span $(0,\infty)$.

We wish to follow the trajectory of a solution of norm $u(0) = A$, hopefully for the corresponding $\lambda(A)$ getting closer and closer to the asymptote $\lambda_\infty = 2$. Let

$$
t = -\ln r - A/2 - \ln(\lambda/2)2
\tag{3.19}
$$

be the above mentioned time-like variable, the translation constants obviously the result of some aforethought, and let

$$
y_1 = u + 2\ln r + \ln(\lambda/2)
\tag{3.20}
$$

be the desired solution u after translation by u_∞ as indicated. Under these change of variables the equation (3.18) takes the 1st order system form

$$
\begin{cases}
\dfrac{dy_1}{dt} = y_2 \\[2em]
\dfrac{dy_2}{dt} = y_2 - 2e^{y_1} + 2 .
\end{cases}
\tag{3.21}
$$

A phase plane trajectory for (3.21) is shown in Figure 6.

176

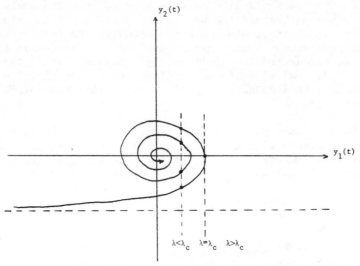

Figure 6.   Phase Plane Trajectory.

The only equilibrium point (r.h.s. of (3.21) = 0) for (3.21) is (0,0) because from $y_2 = 0$ it immediately follows directly from (3.21) that also $y_1 = 0$. At this equilibrium point $u(t) = u_\infty(r)$ and $\lambda = 2$. The initial conditions $u(0) = A$ and $u'(0) = 0$ correspond to trajectories starting in from $(-\infty, -2)$ and wrapping around the origin. The number of solutions to (3.18) may be counted along the vertical dotted lines shown in Figure 6.

4.   LITERATURE, HISTORY, AND SCHOLARSHIP

The ideal research paper should contribute something new to each of the issues (1) physical (2) analytical (3) numerical (4) historical mentioned in the Introduction. Most try to advance knowledge on just one of these issues. Because there is no way to survey here all of the burgeoning literature in this subject, we will be very neglectful of physical issues, purposely limited in our exposition of analytical issues, rather specialized in our numerical survey, and very sparse among references.

We do hope to bring out some not well known historical facts. Some of those will be mentioned in this section, with the next two sections reserved for two important lesser known historical connections.

Both as a vehicle to produce and compare some important and recent references from which the reader may proceed to others, and to support a contention (seldom stated) that historical, taken broadly as scholarship past and present, is in some ways the most difficult component of research, we will proceed by giving a number of literature, history, and scholarship examples that we have personally happened upon recently in connection with these combustion problems. We do this in a spirit of levity, critiqueing ourselves as well as others.

For example, when our paper [1] appeared we were quite surprised to find it flanked on either side by the papers of Zaturska [13] and Lacey and Wake [14] dealing with the same problem. This, in a journal not overly dedicated to combustion problems. Incidentally, we still have not gotten around to reading all of the papers cited in [13] and [14].

As a quick background to further examples of how one keeps up with current literature, last week I was at a conference on computational fluid dynamics (UTSI, Tennessee, 3/16/84). During breaks I sought out the library and happened on a number of examples related to the subject at hand. Let me pass on what I found.

Through no fault of their own, in Roose, Piessens, Hlavacek and Van Rompay [15] the authors conclude that "the technique proposed is apparently the first general procedure for evaluation of critical conditions of explosion and ignition and extinction bounds for 2-D and 3-D cases". However, seven years earlier Anderson and Zienkiewicz [16] presented their general finite element approach with several 2-D and 3-D results. Using finite difference multigrid schemes Chan and Keller [17] did such problems on the unit square, see also Mittelman [18], Meis, Lehmann, and Michael [19]. Earlier there was Simpson [20], Abbott [21], and other analyses mentioned in [17]. Apparently most of these authors were unaware of [16].

In [17] for the unit square λ_c is found by finite dif-ferences to be approximately 6.62 on a coarse (h = 1/3) grid, $\lambda_c \simeq 6.8055$ on a finer (h = 1/29) mesh, whereas $\lambda_c \simeq 6.812$ in [16] by finite elements on a mesh of size (apparently equiva-lent to h = 1/8) intermediate to those just quoted. Further remarks about these mesh dependent solutions will be found in Section 9. To return to the question at hand, how are chemists, engineers of various genera, physicists, and mathe-maticians to keep track of what is being done numerically by the others? Probably the best but far from infallible method is to go to meetings out of ones own discipline. Also, drop in at libraries other than your own when at meetings.

We have just learned (thanks to Joel Dendy, during a brief visit to Los Alamos, 4/3/84) of [17a], in which $\lambda_c \simeq 6.808124$ by an improved [17].

Another paper I found was that of Boddington, Griffiths, and Hasegawa [22]. This interesting work compares experimental, numerical, and theoretical conclusions for time to ignition in supercritical systems. Significant discrepancies are found in the earlier results of Bebernes and Kassoy [23]. Although, as in [22], we don't wish to attempt to resolve those discrepancies here, one may remark that the approach of [23] is heavily based on the earlier analyses of Kaplun [24] and Fujita [25], [26, see in particular Theorem 2.6] wherein the estimates make free use of Jensen's inequality, which could produce unsharp bounds. A difficulty in precision that can occur in using a method of lines for such problems will be mentioned in Section 9. See Lacey [27] for an affirmative resolution of the question raised in [23] as to finite time blowup for $\lambda_c < \lambda \leq \lambda_B$

Theoretical papers can be, roughly, divided into two sets, which can be, again roughly but for simplicity, labelled

(1) Asymptotics theory,
(2) Bifurcation theory.

An alternate labelling would be, again allowing overlap,

(1') Engineers,
(2') Mathematicians

and we hasten to add that these are not quite the same sets. A good recent reference for (1) is the book Buckmaster and Ludford [28]. See also the recent papers Chung and Law [29] and Laseigne and Olmstead [30]. For (2) in a rigorous sense it seems that the two best tools are fixed point theorems, several of which may be applied in somewhat standard ways to demonstrate existence of solutions, and the phase plane approach illustrated in Section 3 for counting solutions. The full qualitative theory for the bifurcation diagrams lags the rather good quantitative results. See as examples Joseph and Lundgren [31] and Shivaji [32]. For a connection between the sets (1) and (2) see Berestycki, Nicolaenko, and Scheurer [33].

The history of the problem

$$-\Delta u = \lambda e^u \qquad (4.1)$$

goes back at least as far as Louville [34] and Poincaré [35]. An excellent account of the subsequent history is given in Davis [36] and we shall not repeat it. We do wish to distinguish the two cases: $\lambda > 0$ and $\lambda < 0$. The former is the one being treated in this paper. Most of the complex function theory analysis of (4.1) by Poincaré, Picard, Bieberbach, and others in that era (see [36] for the references) apparently dealt primarily with the latter. Much of that development was

centered on the theory of Fuchsian functions. A recent review
of the latter is that of Gray [37].

In keeping with the spirit of this section, we must
observe that [37] does not mention [35].

Except for what we wish to present in the next two
sections, we now close this section with the admission that
when we did the work in [1] we were completely unaware of that
of Kubicek and Hlavacek [38]. Our principal motivation was to
analyze in a numerically exact way the differences between the
bifurcation diagrams of the true Arrhenius equation (1), the
Frank-Kamenetskii approximation (4.1), and the Bazley and Wake
[39] approximation

$$-\Delta u = \lambda e^{\dfrac{u}{1+\epsilon\|u\|}} . \tag{4.2}$$

As it turns out our approach was quite similar to that of [38],
more generally called the GPM algorithm, see Kubicek and
Hlavacek [40]. Then we added the higher order calculus HOC
algorithm, to be described in Section 7.

As a final "literature" pique on ourselves, we were mildly
chagrined to see that the last two references of [3], to which
several specific comparisons were made, somehow were dropped
off the bottom of the page [3, p. 868]. They were:

[3, Reference [14]] = Reference [2] of this paper;

[3, Reference [15]] = BODDINGTON, T., FENG, CHANG-GEN, and
GRAY, P., Disappearance of Criticality in Thermal Explosion under Frank-Kamenetskii Boundary Conditions: Comment
on Zaturska and Gill, Donaldson, and Shouman, Combustion
and Flame 48, pp. 303-304, 1982.

5. BRATU AND FRANK-KAMENETSKII

After we began the work in [1] we noticed numerous
references to equation (1), especially in the case $\epsilon = 0$, as
Bratu's equation. Somewhat in the spirit of the historical
problem of "why is it called the Robin boundary condition" of
[5], see also [41], we began asking several who so referred to
the equation, as to the origin of the name, without luck.
Success came simultaneously in summer 1983, from John Rae at
Harwell, who responded to my throwing the question out at the
1983 Seattle Conference, and from Nahil Sobh, who was given the
historical question to answer as part of a summer reading
course. Both directed me to the references to Bratu's work
given in Davis' book [36].

In particular, in considering the equation (1) in one
dimension with $\epsilon = 0$, in [42, section 31], Bratu calculates

λ_c analytically as

$$\lambda = \frac{(1.8745...)^2}{4} = 0.8784 \qquad (5.1)$$

Moreover he demonstrates in [42] that for $0 < \lambda < \lambda_c$ there are two separate solutions which come together to the one at $\lambda = \lambda_c$. His value for λ_c clearly compares favorably to the value $\lambda_c = 0.878$ calculated later by Frank-Kamenetskii [43].

Bratu announced his results without proof and with less numerical precision earlier in [44]. The value of λ_c is found there to be

$$\lambda_c = \frac{2}{2.27} = 0.88 \qquad (5.2)$$

to the precision given. Without going into all the details, it is interesting to note how Bratu calculates λ_c. Let τ denote the real root of the equation

$$\log(\sqrt{t} + \sqrt{t-1}) = \sqrt{\frac{t}{t-1}}. \qquad (5.3)$$

and let

$$\beta = 2 \sqrt{\frac{2}{\lambda\tau}} \log(\sqrt{\tau} + \sqrt{\tau-1} \qquad (5.4)$$

$$= \frac{2\sqrt{2}}{\sqrt{\lambda} \cdot \sqrt{\tau-1}}$$

Letting $h = \beta\sqrt{\lambda}$, then

$$\lambda_c = \frac{h^2}{4} = \frac{2}{\tau-1}. \qquad (5.5)$$

In [44] $\tau \simeq 3.27$ with higher precision given in [42]. The precision in λ_c is limited only by the root-finding procedure used for equation (5.3).

Equation (1) has been called Bratu's equation, the Poisson-Boltzman equation, Frank-Kamenetskii's equation, and Liouville's equation. As we will see in the next section, it could also be called Emden's Equation or Chandrasekhar's Equation.

6. CHANDRASEKHAR AND STELLAR THEORY

A main idea of Bazley and Wake [45] is to use the
Chandrasekhar and Wares [46,47] numerically tabulated values of
the isothermal functions of stellar theory to resolve the
apprpoximate equation (4.2) in three dimensions. This idea was
used earlier by Chambre [48] for the Frank-Kamenetskii case
(i.e., $\varepsilon = 0$). The book of Davis [36] gives a good exposition
of the Emden and Chandrasekhar theory of such equations.

A main idea of [1] was to calculate directly the thermal
functions of equation (1) without taking recourse to any other
calculated values. The premise behind this idea is simply that
the mathematicians' ODE solvers of today are much better than
before. However to get very high precision values of the
critical parameters we had to develop a higher order calculus
(HOC) algorithm.

One naturally wonders therefore if the numerical state of
the art as carried out by the stellar theorists had not also
advanced. We have looked into this, but only briefly so our
impressions are vulnerable. It seems that the numerical
modelling in the stellar theory has moved away from the
relatively theoretical approach taken by Chandrasekhar. It has
instead been strongly influenced by empirical models heavily
influenced by observational data. Many effects, primary and
secondary, such as convective effects, mixing processes, mass
loss via winds, turbulent diffusion, magnetic field effects,
are taken into account in these models. A good reference is
Iben [49].

This is not to say that the combustion~stellar analogy
should not be pursued. Indeed a similar type of modeling of
multiconstituent combustion processes has built up in combustion theory. See for example the work of the Sandia group,
such as Margolis and Matkowsky [50]. We are not prepared to
discuss those models here. Their numerical calculation uses
PDE codes such as PDECOL and other general software for time
dependent, espeically parabolic, systems.

7. ELABORATION ON NUMERICAL METHODS

As set out in Section 1, we group the numerical approaches
as those using

(1) Shooting or similar techniques.
(2) Linearization or similar techniques.
(3) Continuation or similar techniques.

We have already mentioned a number of the principal papers. In particular let us point out the recent Kubicek-Hlavacek book [40] for (1), a recent paper by Boddington, Feng, and Gray [51] for (2), and Chan and Keller [17] for (3). From these one has a quick gateway to the essentials of the three principal approaches and to relevant recent literature.

We should make clear that we will miss many papers on numerical methods for these problems. We can only report those that we know. For example there are a great many results appearing from what we think of as the English (e.g., Gray et al.) school, the New Zealand (e.g., Wake et al.) school, the Polish (e.g., Kordylewski et al.) school, among others. One of our goals in this paper has been to present not only the main trends but also some lesser known important results.

For example, consider a categorization of combustion numerical methods by

    (a) Finite difference methods,
    (b) Finite element methods,
    (c) Spectral methods.

Most are (a). As far as we can recollect at this moment, we have not mentioned any of (c), although almost surely they exist.

Consider then (b) and in particular the work of Anderson and Zienkiewicz [16]. Note that [16] includes

    (i) computed critical parameters (in the Frank-Kamenetskii approximation) for a variety of domains including ellipses and more complex composite geometries.

    (ii) a clear demonstration (albeit numerical) that on nonsymmetric domains, e.g., ellipse, the solutions are not symmetric (granted, they have some kind of group property), in contradistinction to the situation of Gidas et al. [9].

    (iii) calculations for finite Biot number.

Indeed by the FEM (6) one should be able to consider rather arbitrary constituent configurations. This could be quite important when the process of assembly before reaction is also included.

We turn to another matter. Often one sees the basic paper Kordylewski [52] listed as a beginning point of approach (2). Sometimes this is stated as the starting point of bifurcation interpretations of thermal events. But Luss and Amundson [53] earlier used bifurcation theory to show the stability of chemical systems with a unique steady state.

We conclude this section with a brief summary of the three numerical approaches discussed here. For more details see [40], [51], [17] and the citations therein.

Shooting (1) is a familiar technique. For combustion applications it is usually better to specify $u(0) = A$ rather than the derivative $u'(0) = \alpha$ commonly used, for the simple reason that most of these combustion problems are analyzed on spherical domains and one shoots from the center value where α is always zero. Due to the positivity of all solutions, there are no intermediate node problems. There are however stiff and large argument problems. Our particular method is detailed in the next section. Attempts to do shooting on higher dimensional nonspherical domains will be described in Section 9.

Linearization (2) proceeds as follows. Let u be the solution to a nonlinear problem

$$\Delta u + \lambda F(u) = 0 \quad \text{in } r < 1 \text{ in } R^n,$$
$$u = 0 \quad \text{on } r = 1,$$

(7.1)

which by symmetry reduces to

$$\frac{1}{r^{n-1}} \frac{d}{dr}\left(r^{n-1} \frac{du}{dr}\right) + \lambda F(u) = 0 \quad \text{in } 0 < r < 1,$$
$$u'(0) = u(1) = 0.$$

(7.2)

Assume $u > 0$ $F(u) > 0$ so that there is only one maximum value of the solution. Consider for comparison the linear problem

$$\frac{1}{r^{n-1}} \frac{d}{dr}\left(r^{n-1} \frac{d\phi}{dr}\right) + \lambda \phi = 0 \quad \text{in } 0 < r < 1,$$
$$\phi'(0) = \phi(1) = 0.$$

(7.3)

Let ϕ_1 and λ_1 be the first eigenfunction and eigenvalue, respectively. Recall that ϕ_1 is positive in $0 < r < 1$. Integrate by parts

$$\int_0^1 \frac{du}{dr} \frac{d\phi_1}{dr} r^{n-1} dr = \lambda_1 \int_0^1 u\phi_1 r^{n-1} dr = \lambda_1 \int_0^1 F(u(r))\phi_1(r) r^{n-1} dr.$$

(7.4)

Therefore λ is characterized variationally by

$$\lambda = \lambda_1 \frac{\int_0^1 u\phi_1 r^{n-1} dr}{\int_0^1 F(u)\phi_1 r^{n-1} dr} . \qquad (7.5)$$

Assume now $F(u)/u$ is bounded below by some $m \geq 0$. For example for the Frank-Kamenetskii case $F(u) = e^u$ one has

$$F(s)/s = \frac{1}{s} + 1 + \frac{s}{2!} + \frac{s^2}{3!} + \dots \geq 1 . \qquad (7.6)$$

Substituting $F(u)/u \geq m$ into (7.5) provides an upper bound for λ for which solutions u can exist, the least such value being λ_c . The numerical calculation of u and λ follow by (a) further sharper variational characterizations involving the functional derivatives F_u and F_{uu} , and then

(b) a time dependent scheme for the calculation of the temperature, linearized eigenfunction, and eigenvalue. This part (b) of the numerical part of approach (2) could be viewed then as a pseudo unsteady (i.e., physically, false transient) method although such a view is not completely correct inasmuch as the time-dependent process can have some real significance.

Arc length continuation methods (3) are variations on Newton's methods usually using a prediction-correction scheme. Given a value (u_n, λ_n) numerically on the bifurcation curve, one

   (i) computes the derivative u_λ^n by the functional linearization of the problem at (u_n, λ) ,

   (ii) does a linear predictor estimate to $\tilde{u}_n$,

   (iii) Newton iterates $\tilde{u}_n^{(k)}$ to correct λ to convergence.

If the bifurcation curve has a vertical slope, one uses an old folklore trick (see the discussion in [1]) to turn the graph 90° and reverse the roles of u and λ . For cases where this does not work, parametrize in terms of arc length rather than by $u(\lambda)$ or $\lambda(u)$. See [17] for more on approach (3). A variation on this approach may be found in [18].

8.   THE HIGHER ORDER CALCULUS (HOC) SCHEME

In [1] we solved and compared the true Arrhenius problem
(1.1), the Frank-Kamenetskii approximation (4.1), and the
Bazky-Wake approximation (4.2), using a scheme combining
shooting, a Newton-Raphson technique, and certain boundary
value expansions. Later we found the paper [38] and the book
[40]; the HOC scheme of [1,2] may be regarded as an extension
of the GPM schemes of [38,40]. Our algorithms enabled us to
obtain full bifurcation diagrams, solution profiles, and the
critical ignition parameters in all cases to effectively
unlimited accuracy. By the seemingly innocuous device of
specifying $\|u\|$ and solving for λ (the old folklore trick
mentioned just above), rather than the more common current
habit of specifying λ and solving for $\|u\|$, we avoided
several numerical difficulties near branching points.

The radial form of the Arrhenius equation (and the others
as well, with obvious modifications) under the Dirichlet
boundary condition (4) and the known solution symmetry becomes,
as in (1.2), (3.18), and (7.2)

$$
\begin{cases}
\dfrac{d^2u}{dr^2} + \dfrac{j}{r}\dfrac{du}{dr} + \lambda \, \exp(u/(1+\varepsilon u)) = 0 \ , \quad 0 \le r < 1 \ , \\[2mm]
u'(0) = u(1) = 0 \ ,
\end{cases}
\tag{8.1}
$$

where $j = 0$ for the infinite slab geometry, $j = 1$ for the
infinite cylinder geometry, $j = 2$ for the sphere in three
dimensions. Letting $y_1 = u$, $y_1 = u' = y_1'$, equation (8.1)
is equivalent to the first order system

$$
\begin{cases}
y_1' = y_2 \\[1mm]
y_2' = -(j/r)y_2 - \lambda \, \exp(y_1/(1+\varepsilon y_1)) \\[1mm]
y_1(0) = A = \|u\| \\[1mm]
y_2(0) = 0 \ .
\end{cases}
\tag{8.2}
$$

Assuming that the solution is a continuously differentiable
function of λ , we may write the Taylor series expansion of
y_1 evaluated at the endpoint $r = 1$, truncated after the
linear term,

$$
y_1(1;\lambda^{i+1}) = y_1(1;\lambda^i) + (\lambda^{i+1}-\lambda^i)\, \frac{dy_1(1)}{d\lambda}\Big|_{\lambda^i} \ .
\tag{8.3}
$$

To obtain the desired endpoint $y_1(1;\lambda^{i+1}) = 0$, we adjust λ
by the Newton-Raphson technique

186

$$\lambda^{i+1} = \lambda^i - \frac{y_1(1,\lambda^i)}{y_{1,\lambda}(1;\lambda^i)} \tag{8.4}$$

using the notation $y_{1,\lambda} = dy_1(1)/d\lambda$ evaluated at $\lambda = \lambda^i$.

Finding a value for $y_{1,\lambda}$ is accomplished by integrating auxiliary equations obtained by differentiating the equations for y_1 and y_2 with respect to λ . Assuming the solution is twice continuously differentiable in λ , we arrive at, letting $y_3 = y_{1,\lambda}$ and $y_4 = y_{2,\lambda}$, the auxiliary equations

$$\begin{cases} y_3' = y_4 \\ y_4' = -(j/r)y_4 - \left[1 + \lambda(y_3(1+\varepsilon y_1))^2\right]\exp(y_1/(1+\varepsilon y_1)) \\ y_3(0) = 0 \\ y_4(0) = 0 \ . \end{cases} \tag{8.5}$$

The system (8.2), (8.5) was solved using a "Gear" package (multistep, variable stepsize, variable order, Adams method). This provided the needed values $y_1(1;\lambda)$ and $y_{1,\lambda}(1,\lambda)$ for the Newton-Raphson iterations.

To obtain efficient exact calculation of the critical inflection point $\lambda_c(\varepsilon_c)$, see Figure 1, we then needed to extend the method to a higher order numerical calculus in the two parameters λ and ε , rather than in just one parameter λ as was done in [38]. We have detailed the features of this extension in [2], where we have also indicated its use for general two parameter ordinary differential equations

$$y'' = f(x,y,y',\lambda,\varepsilon) \tag{8.6}$$

under general boundary conditions. The exact calculation of the critical branching point is not trivial because one does not have an explicit representation of λ as a function of $\|u\|$ and ε . In principle the method of [2] extends to multiple primary (e.g., $\lambda_1,\ldots,\lambda_K$) and secondary (e.g., $\varepsilon_1,\ldots,\varepsilon_\ell$) parameters, so long as the associated initial value problems exhibit numerical stability, and provided one is willing to work out the needed multiparameter calculus.

See [2,4] for full information on the additional equations and details of the HOC scheme. In particular it is easily modified [4] to yield exact values of all secondary ignition and extinction parameters. See the next section for a preliminary discussion of that problem.

Here are examples of the convergence history typically obtained for the exact criticality calculation. For the full Arrhenius equation for cylinder geometry we began with a ballpark guess $u(0) = 5$, $\lambda = 1.5$, $\varepsilon = 0.25$, the iteration then progressing as shown in Table 2.

$u(0)$	λ	ε
5.000000	1.500000	0.250000
4.510696	2.519950	0.224926
5.086746	3.028820	0.245826
5.842395	3.030639	0.243342
5.943262	3.006342	0.242102
5.943243	3.006301	0.242106

TABLE 2.   Iteration History for Cylinder.

For the sphere we started with $u(0) = 7$, $\lambda = 5$, $\varepsilon = 0.24$, the ensuing convergence is given in Table 3.

$u(0)$	λ	ε
7.000000	5.000000	0.240000
7.167618	5.043875	0.238801
7.182183	5.041864	0.238882
7.184861	5.041087	0.238796
7.184943	5.041112	0.238797

TABLE 3.   Iteration History for Sphere.

We believe that the critical ignition parameters so obtained are accurate to six significant digits and at the very least improve all others previously found. See however the recent results obtained, tabulated, and discussed in [51]. Although in the combustion problem under treatment higher accuracy may not be needed, essentially unlimited accuracy in similar problems can be obtained by the method of [2]. Such might be useful in treating criticality calculations in more complicated multicomponent multiphase problems.

9.   FURTHER DISCUSSION OF NUMERICAL MATTERS

In this section we give a number of informal remarks, observations, admissions, and references of possible interest to particular readers.

Neither we [1,2,3] in our HOC scheme nor [38,40] in their GPM scheme pay much explicit attention to the stiffness aspects of the equations being modelled. In particular we are able to push our algorithm to failure by reducing ε to $\varepsilon = 0.0001$ and trying to continue in $\|u\|$ for relatively large $\|u\|$. In that case one encounters very large nonlinearity values. Currently we are implementing LSODI for use in HOC and for experimentation concerning stiffness and the handling of large nonlinearities.

See the papers by Dahlquist, Ames, Gear, and others in Hinze [54] for analyses of the problems of stiffness occurring in chemical kinetics. Stiffness also can damage attempts to use ODE solvers in Method of Lines (MOL) schemes for time dependent integrations used in combustion calculations. In particular, it can incorrectly indicate a "time of blowup". Examples of this using a number of ODE solvers will be found in [4].

We investigated the use of MOL in conjunction with shooting for the combustion equations on rectangular domains, as an alternative to the arc-length continuation schemes. Results will be found in [4]. As was mentioned in [15] one can expect a system of stiff equations. For elliptic equations such an approach was investigated earlier by Jones, South, and Klunker [55]. Roughly, the more lines used, the more unstable the method becomes for elliptic problems.

At the singularity $r = 0$ in (8.1) one should, strictly, solve the modified equation

$$
\begin{cases}
\dfrac{d^2 u}{dr^2} + \dfrac{\lambda}{1+j} \, \exp(u/1+\varepsilon u) = 0 \\[2mm]
u'(0) = u(1) = 0
\end{cases}
\tag{9.1}
$$

for the first step when shooting from $u(0) = A$. Instead we employed an ad hoc ansatz that $y_2 j/r = 0$ for the first step. We ran it both ways (modified first step, unmodified first step) and found no significant differences. At first this may seem surprising. For example we tested the equation (4.2) for the sphere geometry with $\varepsilon = 0.2$, $A = 20$, and found values for $(j/r)(du/dr)$ of -24.9, -410.9, -411.5, -411.4666 for $r = 1.0, 0.01, 0.0001,$ and 0.000001, respectively. The theoretically correct limit is

$$
\lim_{r \to 0} \frac{j}{r} \frac{du}{dr} = u''(0) .
\tag{9.2}
$$

The reason the computation is insensitive to the first step ansatz is apparently because the first step was $0(10^{-8})$.

In comparing approaches (1) and (2), note that the expression (following the exposition of [51]) $f_{\theta\theta}$ has the form, for the Arrhenius exponent ,

$$
f_{\theta\theta}(\theta) = \frac{e^{\frac{\theta}{1+\varepsilon\theta}}}{(1+\varepsilon\theta)^4} \left[1 - 2\varepsilon - 2\varepsilon^2 \theta\right] .
\tag{9.3}
$$

The same expression occurs in our [1, eqn. (14c)]. In this way
one can see that the approaches (1) and (2) are perhaps not so
fundamentally different.

Thus far, arc length continuation approaches (3) do not
appear to have employed as high an order of refinement. In
particular, apparently the bifurcation diagrams found there are
rather mesh dependent and will shift position according to the
mesh used. In some applications where the true bifurcation
diagram is not qualitatively known, one cannot tell the
spurious diagrams from the real one.

To obtain some idea as to the precision possible under
approach (1) for calculating complicated bifurcation diagrams,
we recently ran both the Arrhenius equation and the approximation (4.2) with $\varepsilon = 0.01$ for the sphere geometry for the
full range $\|u\| = 0$ to $\|u\| = 100$, and (roughly) plotted the
results shown in Figure 7. The output appears to be good to 3
or 4 decimals, our plotting purposely somewhat distorted in
Figure 7 to enhance the apparent relative maxima and minima
(ignitions and extinctions). An exact calculation of the
number of such, using the HOC scheme, will be found in [4]. It
is clear from Figure 7 that there is no hope of expecting good
qualitative approximation of the Arrhenius solution from that
of (4.2) after the fourth branch.

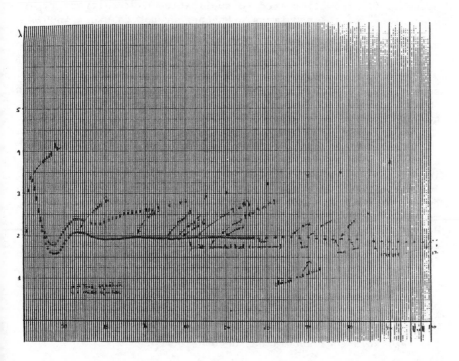

Figure 7. Multiple Ignition and Extinction Curve.

We close this section by pointing out some recent important studies bearing on thse combustion problems, with somewhat different aspects in mind.

An excellent discussion of turbulence modelling for combustion processes may be found in Ghoniem, Chorin, and Oppenheim [56]. The random vortex model is seen to give remarkably good simulations in the cases reported.

Lee and Ramos [57] compare finite element and finite difference simulations of an integrodifferential model for a one-dimensional traveling wave solution to flame propagation. A difference of less than 2% is observed. In [57] a Lewis number = 1 is used.

In the workshop [58] Peters and Warnatz report on the considerable difficulties encountered when running flame propagation codes on test problems with Lewis number $\neq 1$. At Lewis number $Le = 2$ and activation energy $\beta = 20$ only one of five codes could resolve the problem.

10. FUTURE TRENDS

Reactant consumption and reaction byproducts are now being treated by these methods. In particular the ODE solvers approach (1) is well suited to application to "arbitrary right hand sides" of the equations. Whether or not to use a (HOC) scheme or just let the solver do a default jacobian will depend on the precision desired.

Finite element methods will become more important especially for handling and studying the effects of geometry on the combustion process. It is harder to judge at this point how spectral methods will come in.

A third trend will be the need to get more of the Navier-Stokes equations into the analysis. An example of this trend is the paper by Kordylewski [59]. The coupling of the Euler, full-potential, and thin Navier-Stokes codes recently developed by the aeronautics industry, with those describing the elementary reactions, rate coefficients, and concentrations, as described for example in Gardiner [60], should provide significant advances. In this way the fluid dynamics and gas dynamics can be brought into the combustion analysis.

The quantitative (i.e., numerical) results should be used to advance the qualitative (e.g., bifurcation) understanding. Along with this the forgotten facts (e.g., as found in Bratu's work) concerning full complex solutions and how they bear on the real solutions should not be overlooked.

REFERENCES

1. GUSTAFSON, K. and EATON, B. - Exact Solutions and Ignition Parameters in the Arrhenius Conduction Theory of Gaseous Thermal Explosion. Zeit. fur Angewandte Math. und Physik, Vol. 33, pp. 392-405, 1982.

2. EATON, B. and GUSTAFSON, K. - Calculation of Critical Branching Points in Two Parameter Bifurcation Problems. J. Comp. Physics, Vol. 50, pp. 171-177, 1983.

3. GUSTAFSON, K. - Numerical Calculation of Critical Parameters of Thermal Explosion, Numerical Methods in Thermal Problems, Ed. Lewis, R.W., Johnson, J.A., and Smith, W.R., Pineridge Press, 1983.

4. GUSTAFSON, K. - in preparation.

5. GUSTAFSON, K. - Partial Differential Equations, John Wiley and Sons, 1980.

6. AMANN, H. - Fixed Point Equations and Nonlinear Eigenvalue Problems in Ordered Banach Spaces. SIAM Review, Vol. 18, pp. 620-709, 1976.

7. AMANN, H. - Existence and Stability of Solutions for Semi-linear Parabolic Systems and Applications to Some Diffusion Reaction Equations. Proc. Roy. Soc. Edinburgh, Vol. 81A, pp. 35-47, 1978.

8. LAETSCH, T. - The Number of Solutions of a Nonlinear Two Point Boundary Value Problem. Indiana Univ. Math. J., Vol. 20, pp. 1-13, 1970.

9. GIDAS, B., NI, W., and NIRENBERG, L. - Symmetry and Related Properties via the Maximum Principle. Comm. Math. Phys., Vol. 68, pp. 209-243, 1979.

10. KELLER, H., and COHEN, D. - Some Position Problems Suggested by Nonlinear Heat Generation. J. Math. Mech., Vol. 16, pp. 1361-1376, 1967.

11. GELFAND, I. - Some Problems in the Theory of Quasilinear Equations. Amer. Math. Soc. Transl., Vol. 29, pp. 295-381, 1963.

12. EMDEN, V. - Gaskugeln, Leipzig, 1907.

13. ZATURSKA, M. - Approximations in the Thermal Explosion Theory and the Nature of the Degenerate Critical Point. Zeit. fur Angewandte Math. und Physik, Vol. 33, pp. 379-391, 1982.

14. LACEY, A. and WAKE, G. - On the Disappearance of Criticality in the Theory of Thermal Ignition. Zeit. fur Angewandte Math. und Physik, Vol. 33, pp. 406-407, 1982.

15. ROOSE, D., PIESSENS, R., HLAVACEK, V. and Van ROMPAY, P. - Direct Evaluation of Critical Conditions for Thermal Explosion and Catalytic Reaction. Combustion and Flame, Vol. 55, pp. 323-329, 1984.

16. ANDERSON, C. and ZIENKIEWICZ, O. - Spontaneous Ignition: Finite Element Solutions for Steady and Transient Conditions. J. Heat Transfer, August, pp. 398-404, 1974.

17. CHAN, T. and KELLER, H. - Arc-length Continuation and Multi-grid Techniques for Nonlinear Elliptic Eigenvalue Problems. SIAM J. Sci. Stat. Comput., Vol. 3, pp. 173-197, 1982.

17a. BOLSTAD, J. and KELLER, H. - A Multigrid Continuation Method for Elliptic Problems with Turning Points. Preprint, December, 1983.

18. MITTELMAN, H. - Multigrid Methods for Simple Bifurcation Problems. Multigrid Methods, Ed. Hackbush, W. and Trottenberg, U., Springer-Verlag, 1982.

19. MEIS, T., LEHMANN, H. and MICHAEL, H. - Application of the Multigrid Method to a Nonlinear Indefinite Problem. Multigrid Methods, Ed. Hackbush, W. and Trottenberg, U., Springer-Verlag, 1982.

20. SIMPSON, R.B. - A Method for the Numerical Determination of Bifurcation States of Nonlinear Systems of Equations. SIAM J. NUM. ANAL., Vol. 12, pp. 439-451, 1975.

21. ABBOTT, J.P. - An Efficient Algorithm for the Determination of Certain Bifurcation Points. J. Comput. Appl. Math., Vol. 4, pp. 19-27, 1978.

22. BODDINGTON, T., GRIFFITHS, J., and HASEGAWA, K. - Induction Times to Thermal Ignition in Systems with Distributed Temperatures: An Experimental Test of Theoretical Interpretation. Combustion and Flame, Vol. 55, pp. 297-305, 1984.

23. BEBERNES, J. and KASSOY, D. - A Mathematical Analysis of Blowup for Thermal Reactions - The Spatially Nonhomogeneous Case. SIAM J. Appl. Math., Vol. 40, pp. 476-484, 1981.

24. KAPLUN, S. - On the Growth of Solutions of Quasi-linear Parabolic Equations. Comm. Pure Appl. Math, Vol. 16, pp. 305-330, 1963.

25. FUJITA, H. - On the Nonlinear Equations $\Delta u + \varepsilon^u = 0$ and $v_t = \Delta v + e^v$. Bull. Amer. Math. Soc., Vol. 75, pp. 132-135, 1969.

26. FUJITA, H. - On Some Nonexistence and Nonuniqueness Theorems for Nonlinear Parabolic Equations. Nonlinear Functional Analysis, Part 1. Ed. Browder, F., Amer. Math. Soc., 1970.

27. LACEY, A. - Mathematical Analysis of Thermal Runaway for Spatially Inhomogeneous Reactions. SIAM J. Appl. Math., Vol. 43, pp. 1350-1366, 1983.

28. BUCKMASTER, J. and LUDFORD, G. - Theory of Laminar Flames, Cambridge Press, 1982.

29. CHUNG, S. and LAW, C. - Structure and Extinction of Convective Diffusion Flames with General Lewis Numbers. Combustion and Flame, Vol. 53, pp. 59-79, 1983.

30. LASSEIGNE, D. and OLMSTEAD, W. - Ignition of a Combustible Solid by Convection Heating. Zeit. fur Angewandte Math. und Physik, Vol. 34, pp. 886-898, 1983.

31. JOSEPH, D. and LUNDGREN, T. - Quasilinear Dirichlet Problems Driven by Positive Sources. Arch. Rat. Mech. Anal., Vol. 9, pp. 241-269, 1973.

32. SHIVAJI, R. - Uniqueness Results for a Class of Positone Problems. Nonlinear Analysis, Vol. 7, pp. 223-230, 1983.

33. BERESTYCKI, H., NICOLAENKO, B., and SCHEURER, B. - Traveling Wave Solutions to Reaction Diffusion Systems Modeling Combustion. Contemporary Mathematics, Vol. 17, pp. 189-207, 1983.

34. LIOUVILLE, J. - Sur l'equation aux differences partieiles: $\partial^2 \log\lambda / \partial u \partial v \pm \lambda/2a^2 = 0$. Jour. de Math., Vol. 18, pp. 71-72, 1853.

35. Poincaré, H. - Les Fonctions fuchsiennes et l'equation $\Delta u = e^u$. Jour. de Math., Vol. 4, pp. 137-230, 1898.

36. DAVIS, H. - Introduction to Nonlinear Differential and Integral Equations, Dover, 1962.

37. GRAY, J. - Fuchs and the Theory of Differential Equations. Bull. Amer. Math. Soc., Vol. 10, pp. 1-26, 1984.

38. KUBICEK, M. and HLAVACEK, V. = Direct Evaluation of Branching Points for Equations Arising in the Theory of Explosives of Solid Explosives. J. Comp. Phys., Vol. 17, pp. 79-86, 1975.

39. BAZLEY, N. and WAKE, G. - The Disappearance of Criticality in the Theory of Thermal Ignition. Zeit. fur Angewandte Math. und Physik, Vol. 29, pp. 971-976, 1978.

40. KUBICEK, M. and HLAVACEK, V. - Numerical Solution of Nonlinear Boundary Value Problems with Applications, Prentice-Hall, 1983.

41. GUSTAFSON, K. - The Robin Boundary Condition. Amer. Math. Soc. Not., Vol. 27, pp. 103, 228, 1979.

42. BRATU, G. - Sur les Equations Integrales Non Lineares. Bull. Soc. Math. de France, Vol. 42, pp. 113-142, 1914.

43. FRANK-KAMENETSKII, D. - Acta phys-chim. URSS, Vol. 10, pp. 365, Zh. Fiz. Khim., 13, pp. 738, 1939.

44. Bratu, G. - Sur Certains Equations Integrales Non Lineares. Comples Rendus, Vol. 150, pp. 896-899, 1910.

45. BAZLEY, N. and WAKE, G. - Criticality in a Model for Thermal Ignition in Three or more Dimensions. Zeit. fur Angewandte Math. und Physik, Vol. 32, pp. 594-601, 1981.

46. CHANDRASEKHAR, S. and WARES, G. - The Isothermal Function. Astrophys. J., Vol. 109, pp. 551-556, 1949.

47. CHANDRASEKHAR, S. - An Introduction to the Study of Stellar Structure, Dover, 1957.

48. CHAMBRE, P. - On the Solution of the Poisson-Boltzmann Equation with Application to the Theory of Thermal Explosions. J. Chem. Phys., Vol. 20, pp. 1795-1797, 1952.

49. IBEN, I. - Post Main Sequence Evolution of Single Stars. Ann. Rev. Astron. Astrophys., Vol. 12, pp. 215-257, 1974.

50. MARGOLIS, S. and MATKOWSKI, B. - Flame Propagation with Multiple Fuels. First International Specialists Meeting of the Combustion Institute, Section Francaise, Combustion Institute, 1982.

51. BODDINGTON, T., FENG, G., and GRAY, P. – Thermal
    Explosions, Criticality, and the Disappearance of
    Criticality in Systems with Distributed Temperatures. I.
    Arbitrary Biot Number and General Reaction-rate Laws.
    Proc. Roy. Soc. London, Vol. A390, pp. 247-264, 1983.

52. KORDYLEWSKI, W. – Critical Parameters of Thermal
    Explosion. Combustion and Flame, Vol. 34, pp. 109-117,
    1979.

53. LUSS, D. and AMUNDSON, N. – Some General Observations on
    Tubular Reactor Stability. Can. J. Chem. Eng., 45, pp.
    341-346, 1967.

54. HINZE, J. – Numerical Integration of Differential
    Equations and Large Linear Systems, Springer Lecture Notes
    in Mathematics 968, 1982.

55. JONES, D., SOUTH, J., and KLUNKER, E. – On the Numerical
    Solution of Elliptic Partial Differential Equations by the
    Method of Lines. J. Comp. Phys., Vol. 9, pp. 496-527,
    1972.

56. GHONIEM, A., CHORIN, A., and OPPENHEIM, A. – Numerical
    Modelling of Turbulent Flow in a Combustion Tunnel. Phil.
    Trans. Roy. Soc. London, Vol. A304, pp. 303-325, 1982.

57. LEE, D. and RAMOS, J. – Application of the Finite-Element
    Method to One-Dimensional Flame Propagation Problems.
    AIAA Journal, Vol. 21, pp. 262-269.

58. PETERS, N. and WARNATZ, J. – Numerical Methods in Laminar
    Flame Propagation. Notes on Numerical Fluid Dynamics 6,
    F. Vieweg. and Sohn, 1982.

59. KORDYLEWSKI, W. – Influence of Aerodynamics on the
    Critical Parameters of Thermal Ignition. Intern. J. Num.
    Meth. Engin., Vol. 17, pp. 1081-1091, 1981.

60. GARDINER, W. – The Chemistry of Flames. Scientific
    American, Vol. 246, pp. 110-124, 1982.

CHAPTER 8

THE ANALYSIS OF FREE AND FORCED CONVECTIVE FLOW IN PRIMITIVE
VARIABLES USING THE FINITE ELEMENT METHOD.

A.G. PRASSAS (i), D. HITCHINGS (ii), and M. EL-NAGDY (iii)

(i) Technical Analyst, Management Services Department,
Babcock Power Limited, UK. (ii) Lecturer at Imperial
College, University of London, UK.· (iii) Principal Engineer,
Nuclear Engineering, Babcock Power Limited, UK.

1. INTRODUCTION

For the past few decades the finite difference method has
been paramount for the numerical analysis of problems in
fluid mechanics and heat transfer. Other methods have
been developed, and in recent years the finite element
method has become increasingly important. It was first
used in the solution of field problems by Zienkiewicz and
Cheung [1] in 1965. This probably marked the beginning
of finite element applications in heat transfer.
Subsequent development was rapid and finite element
approaches were proposed for a variety of problems
involving transient [2] and steady state energy and/or
mass transport. The finite element method is gradually
replacing the finite difference method in many problem
areas for two main reasons:-

i.   The inherent advantages of finite element modelling
      which include the ease of accommodating complex geo-
      metry, the ability of "naturally" incorporating the
      differential-type boundary conditions. Also the
      speed with which finite element programs can be
      adapted to solve problems of different configuration
      or specification.

ii.   The finite element method is a well-established
      numerical technique used for the majority of
      problems in structural mechanics. This is an
      incentive for engineers to use the method for
      solving flow and thermal problems. Using the same
      method for solving both solid and fluid domains,
      makes the task of treating the interaction between
      the two domains easier.

In the field of heat exchanger and pressure vessel design it is necessary to perform the flow analysis for a component, use the results to determine the temperature distribution, and finally analyse for stress. If the same computer code is used for all aspects of the problem this often leads to a faster, more economical and neater approach to solving engineering design problems.

For steady incompressible media, the early finite element solution of the energy and mass transport equations in two dimensions employed the stream function - vorticity formulation [3, 4 and 5]. Either the Galerkin or a restricted functional is applied to obtain the finite element equations. This approach yields strongly positive definite discrete equations which can be easily handled. The incompressibility constraint is satisfied identically. However, with this procedure the two unknowns (stream function and vorticity) cannot be solved for simultaneously because the vorticity boundary conditions are not known a priori. A segregated iterative solution method must be used with this procedure which is difficult to extend to three-dimensions.

The stream function approach [6], where a biharmonic equation in stream function represents the governing equations, is widely used in potential problems. The resulting finite element formulation necessitates the use of higher order elements since the continuity of the derivative is required. This adds considerable complications in the element development.

The difficulties associated with the stream function - vorticity and stream function approaches has led to an increasing interest in the solution of the Navier-Stokes and energy equations in terms of the primitive variables of velocity, pressure and temperature. These variables do not pose any problem in applying the boundary conditions and can be easily extended to three dimensions. The equations involved are of the second order for which element slope continuity is not required. The principal difficulty encountered in solving the Navier-Stokes equations using the primitive variables hinges on the imposition of the incompressibility constraint. The finite element modelling of these equations has followed four basic approaches:-

i.   An integrated formulation in which the velocity components and the pressure are solved for simultaneously. This approach is more physical and simple elemental continuity is the minimum requirement, but the resulting discrete equations are non positive definite. The pressure and velocity equations, which can be numerically unstable, are solved simultaneously using partial pivoting [7 and 8]. Although the velocity components calculated using this approach are accurate, the resulting pressure field was in error. The pressure error has subsequently been identified [20] as being due to an incorrect choice of the interpolation polynomial order for pressure. Olson has recently shown [9] that the pressure must be interpolated by a lower order polynomial than that used for the velocity in order to suppress the spurious rigid body modes which would otherwise appear in the pressure field.

ii. Segregated formulations in which the velocity and pressure are uncoupled and solved alternately. This approach has been little explored in finite elements except for Olson's pioneering work [9] and [21].

iii. The solenoidal velocity interpolation method [10] in which the assumed velocity field has zero divergence and hence satisfies the continuity equation exactly. The pressure, which can be eliminated from the equations, is retained in the boundary conditions. This approach however has many difficulties despite its apparent simplicity.

iv. The penalty function method [11-17] in which the incompressibility constraint is imposed through the addition of a penalty function. The main advantage of this method is the elimination of the pressure as a dependent variable and, as with the solenoidal velocity interpolation method, it is only retained in the boundary conditions. The method leads to a set of a strongly positive definite discrete equations. However, it does not work with every element formulation since the shape functions of some elements do not allow them to be incompressible.

A different approach is presented here; namely the selective reduced integration method. This has been successfully used in the solution of incompressible and nearly incompressible orthotropic elasticity problems [18 and 19]. The method used for the orthotropic elasticity can be directly applied to solve the incompressible Stokes flow as shown in the following section.

2. THEORY AND BASIC EQUATIONS

2.1 The Governing Equations

The general equations of convective transfer and diffusion of thermal energy are presented using the primitive variables. The fluid is assumed Newtonian, incompressible, laminar and within the Boussinesq approximation. Variations of all fluid properties are neglected except for the effect of the density on the gravitational force. The effects of viscous dissipation and radiative transport have also been ignored. The geometry of the problem can be arbitrary, both in dimension and shape. Employing the tensor notation and the repeated indices summation convention, the governing equations are expressed as follows:-

a. Conservation of mass

$$u_{i,i} = 0 \tag{1}$$

This equation, also known as the incompressibility constraint (continuity equation), is satisfied when the relative volumetric change (dilitation) vanishes.

b. Conservation of momentum

The motion of Newtonian fluids is governed by the Navier-Stokes equations

$$\rho u_j u_{i,j} - \sigma_{ij,j} + \rho g_i \beta(T-T_o) = 0 \tag{2}$$

where for isotropic fluids the following relations hold:

$$\sigma_{ij} = -p\delta_{ij} + \tau_{ij} \tag{3}$$

$$\tau_{ij} = \lambda e_v \delta_{ij} + 2\mu e_{ij}$$

$$e_{ij} = u(i,j) = \frac{1}{2}\left(\frac{\partial u_i}{\partial x_j} + \frac{\partial u_j}{\partial x_i}\right)$$

$$e_{ii} = u(i,i) = \frac{\partial u_i}{\partial x_i}$$

$$e_v = e_{11} + e_{22} + e_{33}$$

The momentum equations (2) reduce to the Stokes equations for incompressible flow when the convective terms are omitted. The latter equations are linear equations of the parabolic type and provide a good approximation to the solution of equations (2) when Reynolds number is small.

c.   Conservation of energy

The energy equation for the fluid domain can be expressed as:-

$$\rho c_0 u_i T_{,i} + q_{j,j} = S \tag{4}$$

where:

$$q_j = -kT_{,j} \tag{5}$$

S = volumetric heat source term.

In the solid region the energy equation reduces to:

$$q_{j,j} = 0 \tag{6}$$

To complete the above equations the essential and natural boundary conditions are required for the fluid and solid regions of the problem. These conditions are:

i.   For the momentum and continuity equations

$$u_i = f_u \quad \text{on } S_u \tag{7}$$

$$\sigma_i = \sigma_{ij} n_j \quad \text{on } S_\sigma \tag{8}$$

ii.   For the energy equation

$$T = f_T \quad \text{on } S_T \tag{9}$$

$$h = q_i n_i \quad \text{on } S_h \tag{10}$$

2.2   Incompressible Constraint

The stress tensor for isotropic fluids equation (3) can be re-written as:

$$\sigma_{ij} = -p\delta_{ij} + \lambda\delta_{ij} u_{k,k} + 2\mu u_{(i,j)} \tag{11}$$

where μ and λ are the Lame parameters which are independent of the rates of strain, and can be identified as:

$$\mu \quad = \text{dynamic viscosity } (=E/2(1 + n) \text{ for isotropic solids}) \quad\quad (12)$$

$$\lambda \quad = 2\mu n/(1 - 2n) \quad\quad\quad\quad\quad\quad\quad\quad\quad (13)$$

$$E \quad = \text{Young's modulus}$$

$$n \quad = \text{Poisson ratio}$$

The stress tensor (11) can become incompressible by letting the Poisson ratio tend towards 0.5. For an arbitrary stress state the dilitation or unit volume change is given by:

$$e \quad = u_{i,i} \quad \frac{\sigma_{ii}}{3k} \quad\quad\quad\quad\quad\quad (14)$$

where:

$$k \quad = E/3(1-2n) = 2\mu(1-n)/3(1-2n) \quad\quad (15)$$

So for $n \to 0.5$ the dilatation $u_{i,i}$ i.e. the relative volume change vanishes. Therefore as $n \to 0.5$ this characterizes the incompressible stress tensor.

For incompressible fluids, the dilatation is zero and this can be satisfied when n (the Poisson ratio) is made to approach the value 0.5. The stress tensor (11) with λ calculated when $n = 0.5$ will satisfy the incompressibility constraint.

To solve equation (2) with the stress tensor (11) using the finite element method an equivalent variational equation can be obtained using the Galerkin formulation. Physically, this means that equilibrium of the virtual work is achieved by applying the virtual velocity to the terms constituting equation (2). However, the pressure will only give rise to virtual work if volumetric changes occur and since the fluid considered is incompressible then this implies that the pressure's contribution to virtual work is zero. The pressure p can therefore be removed from the stress tensor (11) when the finite element method is used to solve the governing equations. The stress tensor becomes:

$$\sigma_{ij} \quad = \lambda\delta_{ij}u_{k,k} + 2\mu u_{(i,j)} \quad\quad\quad\quad (16)$$

The above analysis shows that equation (2) with the modified stress tensor (16) and $\eta \rightarrow 0.5$ are approximately equivalent to equations (2) and (1) and stress tensor (11) when the solution is obtained using the Galerkin formulation and finite element method.

The main drawback, in letting $\eta \rightarrow 0.5$ in order to make the medium incompressible, is that $\lambda \rightarrow \infty$ in equation (16). This problem, however, can be overcome by using the selective reduced integration method described later in Section 3.

The advantages of imposing the incompressibility constraint in this manner are:

i. Only one equation is needed to solve for the velocity field and hence there are fewer equations to be solved. The pressure is not calculated, but can be recovered by calculating the stress using the computed velocity field. For a three dimensional field the pressure is:

$$p = (\sigma_{11} + \sigma_{22} + \sigma_{33})/3 \qquad (17)$$

ii. The resulting finite element equations are strongly positive definite and standard solution techniques can be used to obtain the solution velocity vector.

2.3 The Normalised Equations

For ease of comparison of results with other investigators, the governing equations are non-dimensionalised in two forms. In one form the velocity is normalised with respect to the thermal diffusivity "α" and in the other form the velocity is referred to the kinematic viscosity "ν". The dimensionless variables are defined as:

$x_i = x_i/d, \; u_i = u_i d/a$
and
$\theta = (T - T_{ref})/(T_h - T_c)$

where d, Th and Tc are the reference length and the hot and cold temperatures respectively.

For $a = \alpha$ the governing equations in the normalised form are:

Equation of conservation of momentum:

$$\frac{1}{Pr} u_j u_{i,j} - \frac{\partial}{\partial x_j} \left(\frac{2\eta}{1-2\eta} u_{k,k} \delta_{ij} + 2u(i,j) \right) + Ra\theta = 0$$

$$(18)$$

Equation of conservation of energy:

$$u_i \frac{\partial\theta}{\partial x_i} - \frac{\partial^2\theta}{\partial x_j \partial x_j} = S \tag{19}$$

The other form of the normalised equations when $a = \nu$ is:

Equation of conservation of momentum:

$$u_j u_{i,j} - \frac{\partial}{\partial x_j} \left(\frac{2\eta}{1-2\eta} u_{k,k} \delta_{ij} + 2u(i,j) \right) + \frac{Ra}{Pr} \theta = 0$$

$$(20)$$

Equation of conservation of energy:

$$u_i \frac{\partial\theta}{\partial x_i} - \frac{1}{Pr} \frac{\partial^2\theta}{\partial x_j \partial x_j} = S \tag{21}$$

3. FINITE ELEMENT EQUATIONS

3.1 Galerkin Formulation

To apply the finite element method, approximate variational equations corresponding to equation (18), (19) or (20), (21) are required which are valid for the whole of the domain under consideration. They can be obtained by using the Galerkin method [22].

a. Equation of Conservation of Momentum

For the equation of conservation of momentum assume u_i^* to be weighting function, the value of which is arbitrary everywhere on the domain except on the boundary S_u. On S_u the velocity is a known boundary condition, and there u_i^* is assumed zero. Rewriting equation (18) or (20).

$$Qu_j u_{i,j} - \sigma_{ij,j} + P\theta = 0 \tag{22}$$

where

$Q \quad = 1$ or $1/Pr$

$P \quad = Ra$ or Ra/Pr $\hfill (23)$

for normalisations for $a = \alpha$ or $a = \nu$ respectively. Apply the Galerkin weighted residual form:

$$\int u_i{}^* Q u_j u_{i,j} dv - \int u_i{}^* \sigma_{ij,j} dv + \int u_i{}^* P \theta dv = 0$$

$$\hfill (24)$$

Using the Green's theorem on the second term of equation (24) to reduce the order of differention and letting:

$$u_i \quad = \phi^t U_i$$

$$u_i{}^* \quad = \phi^t U_i{}^*$$

$$\theta \quad = \phi^t T$$

where

$\Phi \qquad$ - interpolation functions

$U_i \qquad$ - vector of nodal velocities

$T \qquad$ - vector of nodal temperatures

$U_i{}^* \quad$ - vector of nodal velocities of the weighting functions $u_i{}^*$. They are arbitrary except at the boundary S_u where they become zero.

These can then be used to obtain the following matrix equations (for simplicity only the two dimensional equations are shown below).

$$\int \begin{bmatrix} U_1{}^* t & 0 \\ 0 & U_2{}^* t \end{bmatrix} \phi \begin{bmatrix} \phi^t U_1 & \phi^t U_1 \\ \phi^t U_2 & \phi^t U_2 \end{bmatrix}$$

$$\begin{bmatrix} \dfrac{\partial \phi^t}{\partial x_1} & 0 \\ 0 & \dfrac{\partial \phi^t}{\partial x_2} \end{bmatrix} \begin{bmatrix} U_1 \\ U_2 \end{bmatrix} \quad Qdv \quad +$$

$$\int \begin{bmatrix} U_1^{*t} & 0 \\ 0 & U_2^{*t} \end{bmatrix} \begin{bmatrix} \dfrac{\partial \phi^t}{\partial x_1} & 0 & \dfrac{\partial \phi^t}{\partial x_2} \\ 0 & \dfrac{\partial \phi^t}{\partial x_2} & \dfrac{\partial \phi^t}{\partial x_1} \end{bmatrix}$$

$$\begin{bmatrix} 2\kappa(1-\eta) & 2\kappa\eta & 0 \\ 2\kappa\eta & 2\kappa(1-\eta) & 0 \\ 0 & 0 & 1 \end{bmatrix}$$

$$\begin{bmatrix} \dfrac{\partial \phi^t}{\partial x_1} & 0 \\ 0 & \dfrac{\partial \phi^t}{\partial x_2} \\ \dfrac{\partial \phi^t}{\partial x_2} & \dfrac{\partial \phi^t}{\partial x_1} \end{bmatrix} \begin{bmatrix} U_1 \\ U_2 \end{bmatrix} \quad dv \; =$$

$$\int \begin{bmatrix} U_1^{*t} & 0 \\ 0 & U_2^{*t} \end{bmatrix} \begin{bmatrix} \phi & 0 \\ 0 & \phi \end{bmatrix} \begin{bmatrix} t_1 \\ t_2 \end{bmatrix} ds \; -\int \begin{bmatrix} U_1^{*t} & 0 \\ 0 & U_2^{*t} \end{bmatrix}$$

$$\begin{bmatrix} 0 & \phi \\ \phi & 0 \end{bmatrix} \begin{bmatrix} P_1 & 0 \\ 0 & P_2 \end{bmatrix} \begin{bmatrix} \phi^t & 0 \\ 0 & \phi^t \end{bmatrix} \begin{bmatrix} T \\ T \end{bmatrix} \quad dv \; (25)$$

Where $\kappa = 1/(1-2\eta)$ and P_1, P_2 are the components of P in direction 1 and 2 since P is a function of Ra which is in turn function of g_i.

Since U_1^*, U_2^* are non zero arbitrary weighting functions, the equations can only be satisfied by equating the terms that they multiply giving the set of equations to be solved as:

$$\int \phi_{ab}{}^t dv UQ + \int C^t KC dv U = \int dt ds - \int d^t P dT dv \qquad (26)$$

where

$$c^t = \begin{bmatrix} \dfrac{\partial \Phi}{\partial x_1} & 0 & \dfrac{\partial \Phi}{\partial x_2} \\ 0 & \dfrac{\partial \Phi}{\partial x_2} & \dfrac{\partial \Phi}{\partial x_1} \end{bmatrix}$$

$$d^t = \begin{bmatrix} \Phi & 0 \\ 0 & \Phi \end{bmatrix}$$

$$a = \begin{bmatrix} \Phi^t U_1 & \Phi^t U_1 \\ \Phi^t U_2 & \Phi^t U_2 \end{bmatrix}$$

$$b = \begin{bmatrix} \dfrac{\partial \Phi^t}{\partial x_1} & 0 \\ 0 & \dfrac{\partial \Phi^t}{\partial x_2} \end{bmatrix}$$

$$K = \begin{bmatrix} 2\kappa(1-\eta) & 2\kappa\eta & 0 \\ 2\kappa\eta & 2\kappa(1-\eta) & 0 \\ 0 & 0 & 1 \end{bmatrix}$$

$$U = \begin{bmatrix} U_1 \\ U_2 \end{bmatrix}$$

$$P = \begin{bmatrix} P_1 & 0 \\ 0 & P_2 \end{bmatrix}$$

$$t = \begin{bmatrix} t_1 \\ t_2 \end{bmatrix}$$

b. Equations of Conservation of Energy

A similar procedure is followed for the energy equation. Assume θ^* to be a weighting function, the value of which is arbitrary except on the boundary S_T where they are prescribed. Following the same development of the equations as for the momentum equations the set of equations for a two dimensional domain is given below:

$$\int \Phi \mathbf{e} \mathbf{f}^t dv \mathbf{T} + R \int \mathbf{f} \mathbf{f}^t dv \mathbf{T} = R \int \mathbf{f} \mathbf{n} ds + \int \Phi \mathbf{S} ds \qquad (27)$$

where

$$\mathbf{e} = \begin{bmatrix} \Phi^t \mathbf{U}_1 & \Phi^t \mathbf{U}_2 \end{bmatrix} \qquad R = 1 \text{ or } 1/Pr$$

$$\mathbf{f} = \begin{bmatrix} \dfrac{\partial \Phi}{\partial x_1} & \dfrac{\partial \Phi}{\partial x_2} \end{bmatrix}$$

$$\mathbf{n}^t = \begin{bmatrix} n_1 & n_2 \end{bmatrix}$$

3.2 Assembled Equations

The previous derivation has been concerned with a single finite element and the small portion of the continuum it represents. The discrete representation of the entire continuum region of interest is obtained through an assemblage of elements such that there is inter-element continuity for velocity and temperature. The result of such an assembly process is a system of matrix equations for convective fluid flow. These are given below.

$$\mathbf{C}_m \mathbf{U} + \mathbf{D}_m \mathbf{U} = \mathbf{F}_m \qquad (28)$$

$$\mathbf{C}_e \mathbf{T} + \mathbf{D}_e \mathbf{T} = \mathbf{F}_e \qquad (29)$$

Where the various terms can be identified as:

i. Convective contribution (non linear)

$$\mathbf{C}_m = \sum \int \Phi \mathbf{a} \mathbf{b}^t dv \mathbf{Q} \qquad (30)$$

$$\mathbf{C}_e = \sum \int \Phi \mathbf{e} \mathbf{f} dv \qquad (31)$$

ii. Diffusive contribution (linear)

$$\mathbf{D}_m = \sum \int \mathbf{C} \mathbf{K} \mathbf{C}^t dv \qquad (32)$$

$$\mathbf{D}_e = \sum \int R \mathbf{f} \mathbf{f}^t dv \qquad (33)$$

iii. Forcing functions

$$F_m = \sum \int dt\, ds - \sum \int dP d^t T dv \qquad (34)$$

$$F_e = \sum \int R fn\, ds + \sum \int \Phi s\, dv \qquad (35)$$

Where $\sum$ represents assembly of the individual elemental matrices and 'ne' the number of elements.

3.3 Selective Reduced Integration

It was shown in Section 2.2 that, in order to enforce incompressibility in equation (18) or the equivalent finite element form (26), Poisson's ratio must tend towards 0.5. This, however, will make some of the terms in the material matrix **K** approach infinity. To overcome this problem the selective reduced integration procedure is used [18]. This procedure is based upon the observation that, when the product **C**^{t}**KC** (in equation 26) is performed, only the terms of **C** which give rise to volumetric strain retain the infinite terms of **K**. This led to the separation of **C** into two components, the dilitational (volumetric) and deviatoric (shear) contribution. The volumetric contribution of the product **C**^{t}**KC** is integrated using reduced quadrature thus alleviating 'locking'. In other words the component of the matrix which contains the volumetric terms reduces rank, hence reducing the incompressible constraints applied on the whole of the matrix **C**^{t}**KC** Full integration is employed on the remaining component of the matrix **C**^{t}**KC** to retain the rank of the elemental matrices. To demonstrate this consider the mean strain for a typical node α for a 2D problem which is:

$$\varepsilon_{m\alpha} = \tfrac{1}{2}(\varepsilon_1 + \varepsilon_2) = \tfrac{1}{2}\left(\frac{\partial u_1}{\partial x_1} + \frac{\partial u_2}{\partial x_2}\right) \qquad (36)$$

Using the interpolation for $u\nu$:

$$\varepsilon_{m\alpha} = \frac{1}{2} \begin{bmatrix} \dfrac{\partial \Phi_\alpha}{\partial x_1} & \dfrac{\partial \Phi_\alpha}{\partial x_2} \end{bmatrix} \begin{bmatrix} U_{\alpha 1} \\ U_{\alpha 2} \end{bmatrix} \qquad (37)$$

The typical submatrix for node α of the volumetric contribution of matrix **C** is:

$$C_\alpha^{dil} = \frac{1}{2} \begin{bmatrix} \dfrac{\partial \Phi_\alpha}{\partial x_1} & \dfrac{\partial \Phi_\alpha}{\partial x_2} \\[2ex] \dfrac{\partial \Phi_\alpha}{\partial x_1} & \dfrac{\partial \Phi_\alpha}{\partial x_2} \\[2ex] 0 & 0 \end{bmatrix} \tag{38}$$

The total submatrix for node α is:

$$C_\alpha = \begin{bmatrix} \dfrac{\partial \Phi_\alpha}{\partial x_1} & 0 \\[2ex] 0 & \dfrac{\partial \Phi_\alpha}{\partial x_2} \\[2ex] \dfrac{\partial \Phi_\alpha}{\partial x_2} & \dfrac{\partial \Phi_\alpha}{\partial x_1} \end{bmatrix} \tag{39}$$

The deviatoric term is then equal to:

$$C_\alpha^{dev} = C_\alpha - C_\alpha^{dil} = \begin{bmatrix} \dfrac{1}{2}\dfrac{\partial \Phi_\alpha}{\partial x_1} & -\dfrac{1}{2}\dfrac{\partial \Phi_\alpha}{\partial x_2} \\[2ex] \dfrac{1}{2}\dfrac{\partial \Phi_\alpha}{\partial x_1} & \dfrac{1}{2}\dfrac{\partial \Phi_\alpha}{\partial x_2} \\[2ex] \dfrac{\partial \Phi_\alpha}{\partial x_2} & \dfrac{\partial \Phi_\alpha}{\partial x_1} \end{bmatrix} \tag{40}$$

Using the above expression equation (26) becomes:

$$\int \Phi_{ab}{}^t dv U + \int (C^{dil} + C^{dev})^t K (C^{dil} + C^{dev}) dv U$$

$$= \int dt ds - \int d^t P dT dv \tag{41}$$

Since it can be shown that the terms:

$$\int C^{dil^t} K C^{dev} dv \quad \text{and} \quad \int C^{dev^t} K C^{dil} dv$$

reduce to zero, equation (41) for the whole domain (assembled) can be written symbolically as:-

$$C_m U + \bar{D}^{dil}U + D^{dev}U = F_m \tag{42}$$

Where the bar symbol shows that the matrix has been integrated using reduced quadrature. It can be shown that, because of the reduced integration:

$$D^{dil}_m U = 0$$

if U corresponds to a non zero set of displacements that have no volume change. If U implies a change in volume then:-

$$D^{dil}_m U = F$$

but since the terms in D^{dil}_m are large then the force F must be large to produce a finite U. Since the force F is specified as a finite value then this implies that the corresponding U must have no volume change. In other words the equations of the form:

$$C_m U + D_m^{dev}U = F_m \tag{43}$$

are effectively those solved with $D_m^{dil}U$ acting as a constraint for U to have zero volume change.

3.4 Computer Implementation

The previously described equations were incorporated in a general purpose finite element code FINEL, the structure of which is described in Appendix.

Five different elements were developed, three plane 2-D elements of four, eight, and nine nodes. One 9-noded axisymmetric element and one eight noded 3-D cube element. Although FINEL has automatic element integration facilities for standard elements the momentum equations require selective reduced integration and the energy equation is not symmetric, both of which are non-standard FINEL facilities and hence special integration routines were written for these elements. The selective reduced integration is implemented following the paper by Hughes [18]. For further details of the actual implementation within FINEL see [23].

4. SOLUTION TECHNIQUES

In order to solve equations (42) and (29) which are the discrete equivalents of the conservation of momentum and energy equations, three aspects should be observed. The first is the choice of a solution algorithm to linearise the equation (42) and (29), secondly the choice of the procedure needed to solve the equations simultaneously and finally, the solution technique used to solve the resulting linearised set of finite element equations.

4.1 Linearization of discrete equations

For illustrative purpose consider one of the equations say (42). There are at least four schemes for linearising the above equation.

a. The 'initial velocity method' which can be written symbolically as:

$$D_m U^{n+1} = F - C_m(U^n) U^n \qquad (44)$$

The convective term C_m is evaluated using the velocity solution vector from the most recent iteration and multiplied by the same vector to become a pseudo-load vector to the right hand side of the equation before the next iteration.

b. A variant to the initial velocity method is:

$$(D_m + C_m{}^*(U^n)) U^{n+1} = F_m - C^{**}(U^n) U^n \qquad (45)$$

The convective term is split into its symmetric C^* and antisymmetric C^{**} component keeping the former on the left hand side of the equations.

The previous two methods have the advantage that the left hand side of the equation involves only a symmetric matrix which allows efficient solution techniques to be used. However, both of the above solutions have shown very weak convergence characteristics [24-27].

c. The third scheme is referred to as successive substitution or Picard iteration and can be written as:

$$(D_m + C_m(U^n)) U^{n+1} = F_m \qquad (46)$$

d. Finally, an incremental approach known as the Newton-Raphson can be used. The convective term is multiplied by the unknown velocities and expanded as follows:

$$C_m(U^n) U^n = R(U^{n-1}) + \frac{\partial R}{\partial U} \bigg|_{U^{n-1}} \Delta U^{n-1,n} + \ldots$$

Then, if the expansion is truncated so that only first order terms are retained the linearised system is:-

$$(\mathbf{D_m} + \mathbf{C_m}'(\mathbf{U}^{n-1}))\ \Delta\mathbf{U}^{n-1,n} = \mathbf{F_m} -$$

$$(\mathbf{D_m} + \mathbf{C_m}(\mathbf{U}^{n-1}))\mathbf{U}^{n-1} \qquad\qquad (47)$$

where

$$\mathbf{C}'(\mathbf{U}^{n-1}) = \left.\frac{\partial\mathbf{R}}{\partial\mathbf{U}}\right|_{\mathbf{U}^{n-1}}$$

Equation (47) is solved for the increment $\Delta\mathbf{U}$ which is added to the solution vector $\mathbf{U}$ for the next iteration.

The last two schemes both have non symmetric matrices on the left hand side of the equations which will have to be evaluated at every iteration. Both schemes converge reasonably fast [28] with the fourth scheme being somewhat quicker but more sensitive to the assumed initial solution. In this work the third scheme, the Picard iteration, was used because it was less sensitive to the starting solution and was simple to implement.

4.2 Simultaneous Solution of the Energy and Momentum Equations

In order to obtain the velocity and temperature solution vectors equations (42) and (29) must be solved simultaneously. The two equations may or may not be coupled and this must be anticipated by the solution algorithm. For problems of forced convection with fluid properties independent of temperature the two equations are decoupled. This allows the momentum equation (42) to be solved first which is followed by the solution of the energy equation (29). For problems of natural convection the two equations are directly coupled and they must be solved together. This is achieved by a solution algorithm which alternates between the two solutions and uses the latest available values for the velocity and temperature fields.

4.3 Solution Routine

An out of core modified Cholesky decomposition was used to solve the resulting algebraic equations. The solution routine was adjusted to solve non-symmetric equations.

4.4 Convergence Criteria and Relaxation

In general the first iterate is normally assumed to be a
zero velocity solution. However, at high Rayleigh number
when convergence is difficult to achieve, the solution
for a lower Rayleigh number is normally used for the
starting condition. This arrangement was adequate for
Rayleigh numbers up to 10^5.

The iterative procedure is terminated when the normalised
average between two consecutive iterations is less than a
tolerance i.e. where S^n is a solution vector
(temperature or velocities) n is the iteration level and
m is the number of degrees of freedoms for the present
run.

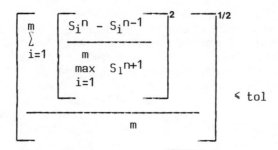

$$\frac{\left[\displaystyle\sum_{i=1}^{m} \left[\frac{S_i{}^n - S_i{}^{n-1}}{\displaystyle\max_{i=1}^{m} S_1{}^{n+1}} \right]^2 \right]^{1/2}}{m} \leqslant \text{tol}$$

In observing the behaviour of the alternating scheme (See
Section 4.2) it was apparent that the large majority of
problems exhibited an oscillatory convergence. To take
advantage of this behaviour and accelerate the
convergence a weighted average of the velocities and
temperature was used in computing the matrix coefficients
for the next iteration. For example the nth iteration

$$U_n = \rho_1 U_n + (1-\rho_1) U_{n-1} \qquad 0 \leqslant \rho_1 \leqslant 1$$
$$T_n = \rho_2 T_n + (1-\rho_2) T_{n-1} \qquad 0 \leqslant \rho_2 \leqslant 1$$

The above arrangement proved to be very adequate and
stable for all problems solved.

5. TEST/APPLICATIONS

In order to demonstrate the validity of the proposed
method and the significance of neglecting the convective
terms in the momentum equation, a series of problems were
analysed. All the numerical tests were performed using
the FINEL general purpose finite element program.

5.1 Significance of Convective Terms

The problem of the square cavity was used to examine the
effect of neglecting the convective terms, since there
are plenty of numerical and experimental results to
compare with. A uniform mesh of 25 elements was used
with the right and left vertical walls being isothermal
with different temperatures (θ_c = -0.5, θ_h = 0.5).
The top and bottom horizontal walls are kept adiabatic.

Results were obtained for Pr = 1 and with different
Rayleigh numbers starting at 10^3 to 10^5.
The results obtained in this investigation were compared
with those obtained by Reddy [14]. These comparisons are
given in Tables 1 and 2 below.

Table 1. Comparison of computation time and number of
iterations required for convergence (n) for the present
investigation and the results of Ref. 14.

Ra	Present Report		Ref. 14	
	n+	cpu*	n+	cpu*
10^3	5	1.14	8	2.33
10^4	6	1.21	16	4.26
10^5	24	3.26	-	-

+   convergence tolerance for both, 10^{-4}.

*   all of the calculations were carried out in double
    precision on an IBM 370/158.

Table 2. Comparison of temperature values obtained by
the present investigation and the results of Ref. 14.

Temperature at Y = 0.5 X Ra	Present Report 10^3	10^4	Ref. 14 10^3	10^4
1.0	0.5	0.5	0.5	0.5
0.9	0.3813	0.2551	0.3871	0.2466
0.8	0.2675	0.07835	0.2777	0.0883
0.7	0.1656	0.00298	0.1765	0.0011
0.6	0.0797	-0.0099	0.0851	-0.0150
0.5	0	0	0	0

As it can be seen from Table 1 the present method converges faster than that of [14] for similar results (see Table 2). Also Figure 1 shows the velocity and temperature distributions at the horizontal mid-plane of the above results at various Ra numbers (10^3 - 10^5).

Further, to substantiate the encouraging results shown the convective terms were calculated during the numerical tests. These terms were compared with the diffusive terms and it was found that for Ra = 10^3 the diffusive terms were at least 100 times larger than the convective terms, and most of them were even larger (i.e. order of 5). For Ra = 10^4 the convective terms were still at least fifteen times smaller than the diffusive terms, with the majority being around forty times smaller.

Only for Ra = 10^5 were some convective terms found to be of the same order as the diffusive ones but, most of the convective terms were small enough to be ignored.

More tests were performed to find out the effect of Prandtl number and it was found that for Pr >1 the convective terms could be ignored for even larger Ra numbers depending on how large the Pr number was. It was also observed that for Pr <1 the solutions were becoming unstable for Rayleigh numbers less than 10^4.

5.2 Natural Convection in Cavities

Further numerical tests were carried out to investigate natural convection in square cavities at various inclinations to the horizontal for Ra numbers of 10^4 and 10^5. A non uniform mesh of 64 elements was used with nodes being closer to the boundary for Ra = 10^5. Figure 2 shows the mesh and the boundary conditions. Figure shows the normalised temperature distribution for Ra numbers of 10^4 and 10^5 at inclinations of -60°, 0°, 60°. As it can be seen from the figure, when the gravity field is inclined towards the cold wall the rate of heat transfer is reduced significantly. This is due to the fact that the buoyancy and convective forces oppose each other. It can be observed for the figure that the opposite happens, when the gravity field is inclined towards the hot wall. The reduction of heat transfer can be seen even more clearly at the velocity vector plots shown in Figure 4. When gravity is inclined towards the cold wall it opposes the flow circulation and it breaks it into cells, the opposite happens when the gravity is inclined towards the hot wall, with the formation of a strong unicellural flow.

Figure 5 shows the average Nusselt Numbers as a function of the angle of inclination for Ra numbers of 10^4 and 10^5. The Nusselt numbers shown here are in close agreement with the reported results of M. Strada and J.C. Heinrich [29].

5.3 Natural Convection in Horizontal Annulus containing Gas enclosed by Solid

Figure 6 shows an annulus containing gas at an atmospheric pressure enclosed between metal pipes lined with insulating material. Using the symmetry argument only half of the pipe cross section was considered. The surface A and B (see Figure 8) kept isothermal at θ_a = 0 and θ_b = 1.0 respectively, surface C was assumed adiabatic. The temperature distribution for both metal/insulation and gas was calculated in a single run. This was achieved by letting velocity field for the solid region be equal to zero. This arrangement is very convenient when the solution technique used deletes the constraint equations from the set of the finite element equations and no extra computations for the momentum equation were needed. The energy equation adjusts itself automatically by ignoring the convective terms for the solid region therefore solving only for conduction. The results shown in Figures 7, 8 are for two different flows of Grashof number (Ra/Pr) of 7.1 x 10^4 and 1.1 x 10^5

5.4 Forced Flow with Heat Transfer

Figures 9 (a and b) show the temperature distributions obtained for a forced flow situation in a mixed domain containing metal and fluid. The same solution procedure as in Section 4.3 is also employed for this problem. The domain is composed of a metal tube of finite wall thickness and length with a linear temperature distribution applied at the outer tube wall with a fully developed parabolic flow at the inlet. A non-uniform mesh was used with 42 nine-noded isoparametric axisymmetric elements for both media (solid and fluid). The results shown in Figure 9b are for similar problems as in Figure 9a except that the axial conduction along the flow direction in the tube is neglected. The second example approximates the case of flow in a thin wall tube.

218

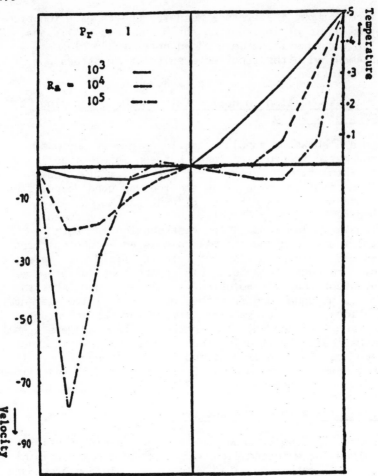

Fig. 1 Velocity and Temperature distribution at
the mid-plane of a square cavity.

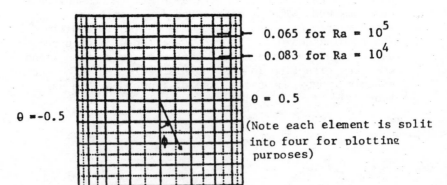

Fig. 2 A non-uniform mesh with 64 elements.

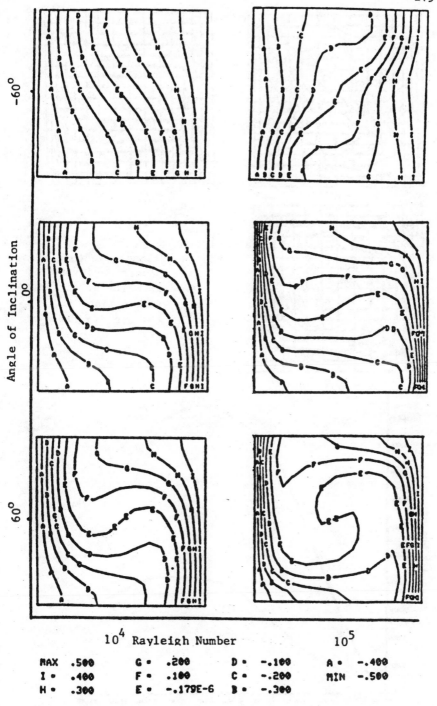

Fig. 3 Temperature isotherms of square cavity for
different Ra numbers and inclinations to gravity
field.

220

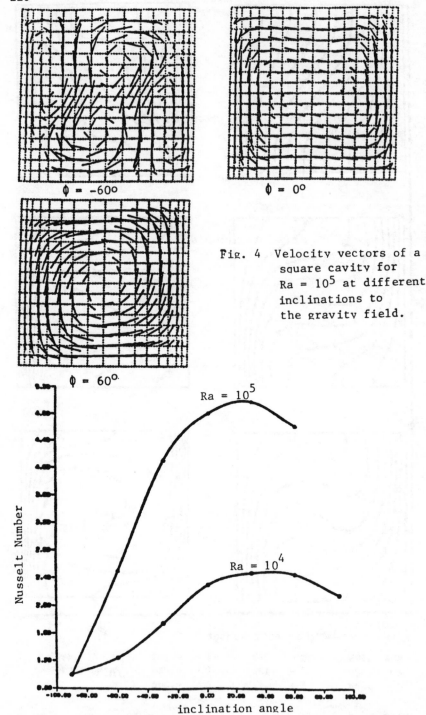

$\phi = -60^\circ$

$\phi = 0^\circ$

Fig. 4 Velocity vectors of a
square cavity for
$Ra = 10^5$ at different
inclinations to
the gravity field.

$\phi = 60^\circ$

Ra = 10^5

Ra = 10^4

Nusselt Number

inclination angle

Fig. 5 Nusselt numbers vs angle of inclination for Ra
numbers of 10^4 and 10^5

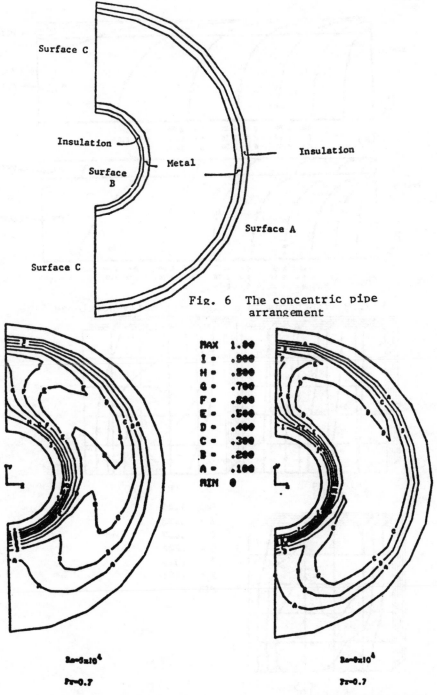

Surface C

Insulation

Surface B

Metal

Insulation

Surface A

Surface C

Fig. 6   The concentric pipe
arrangement

MAX 1.00
I = .900
H = .800
G = .700
F = .600
E = .500
D = .400
C = .300
B = .200
A = .100
MIN 0

$Ra=5x10^4$

$Pr=0.7$

$Ra=9x10^4$

$Pr=0.7$

Fig. 7   Isotherm for the pipe
arrangement shown in
Fig. 6 (Gr = 7.14×10^4)

Fig. 8   Isotherms for the pipe
arrangement shown in
Fig. 6 (Gr = 1.14×10^5)

222

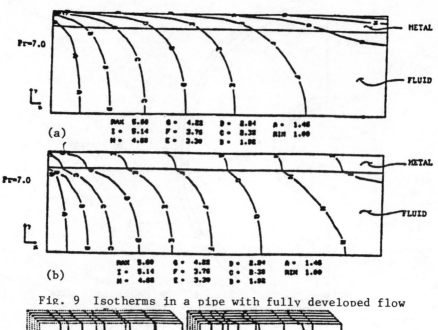

Pr=7.0

METAL

FLUID

(a)

MAX 5.60   G = 4.22   D = 2.84   A = 1.46
I = 5.14   F = 3.75   C = 2.38   MIN 1.00
H = 4.68   E = 3.30   B = 1.92

Pr=7.0

METAL

FLUID

(b)

MAX 5.60   G = 4.22   D = 2.84   A = 1.46
I = 5.14   F = 3.75   C = 2.38   MIN 1.00
H = 4.68   E = 3.30   B = 1.92

Fig. 9   Isotherms in a pipe with fully developed flow

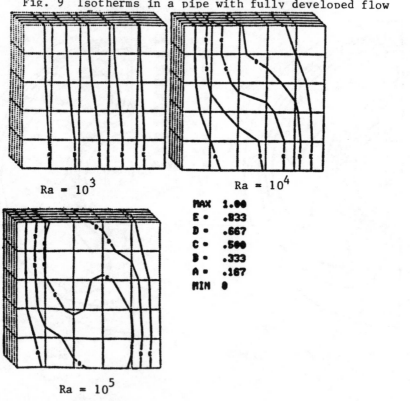

Ra = 10^3 Ra = 10^4

MAX 1.00
E = .833
D = .667
C = .500
B = .333
A = .167
MIN 0

Ra = 10^5

Fig. 10   Isotherm in a cubic enclosure for
different Rayleigh numbers

5.5 Natural Convection in 3-D Enclosures

Finally, some numerical results for a 3-D cube are presented. Figure 10 shows isotherms for Ra = 10^3 to 10^5. The top, bottom, front and back walls, of the enclosure were assumed to be adiabatic while the left and right wall were kept isothermal at two different temperatures ($\theta_c = 0$ and $\theta_h = 1.0$). The velocity field is assumed to be zero at all boundaries.

These results confirm the 2-D assumptions for the similar problem of 2-D square cavity and could be used as test problems for further numerical investigations in 3-D. Some problems with more complicated geometry and boundary conditions were solved using FINEL in Reference [23].

6. CONCLUSIONS

The proposed method has been integrated into a general purpose finite element code FINEL. FINEL was subsequently used to investigate heat transfer by convection for different geometries. The numerical results were compared with published experimental and numerical data.

7. ACKNOWLEDGEMENTS

The authors would like to thank Babcock Power Limited for their financial support and permission to publish the results of this work.

APPENDIX

Structure of FINEL

FINEL is a highly structured program, the basis of which follows directly the steps involved in the finite element method itself. One of the reasons for the success of the finite element method for computer implementation is that it can be broken down into a set of discrete, almost unrelated, steps such as mesh definition, assembly, solution and so on. This feature is used within FINEL to define a modular structure for the program.

Each module within FINEL defines a discrete aspect of the
F.E. Method. The program has a central executive section
which remains resident in core. It is the main purpose of
this executive to control the sequence in which the modules
are executed, and hence to control the analysis that FINEL
carries out. No modules communicate directly with each other,
they all pass through the executive. This is illustrated in
Fig. 1.1.

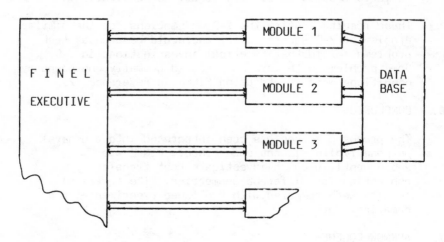

The modular structure is carried down to one lower level
within FINEL by the definition of a series of FINEL
Libraries. Again the idea of these libraries stems naturally
from the finite element method itself. Within the F.E.
analysis a continuum is divided into discrete regions, or
elements. These elements can take different forms depending
upon the actual problem being solved. Within FINEL the
various types of elements available are gathered together
within the FINEL Element Library. Similarly the loadings that
can be applied to the continuum vary according to the problem
being solved. These loading types are then contained within
the FINEL Load Library. There are other libraries within
FINEL, but all of the FINEL Libraries have a similar structure
and serve a similar purpose. Each library has a library
executive program associated with it, and this executive can
be called by any module within FINEL. The library executive
program then calls the actual entry required from the relevant
library.

Finally the input is in the form of data blocks, where each
block has the same general form:

    COMMAND LABEL DATA

The COMMAND is a specific instruction to the program and the LABEL qualifies that instruction. These are usually English words. The DATA is then any other information that the COMMAND requires. It is usually numerical. The execution is controlled by the FINEL executive which calls up each module as required. The modules do not communicate directly with each other: instead, they read and write to the data base. The general form is illustrated in Fig. A1. This structure allows the program to be stopped and restarted between any modules and for the sequence of the module execution to be controlled by the user.

REFERENCES

1.	Zienkiewicz, O.C. and Cheung, Y.K. - Finite Elements in the Solution of Field Problems. The Engineer, Vol. 220 pp. 507-515, 1965.

2.	Wilson, E.L. and Nickell, R.E. - Application of the F.E.M. to the Heat Conduction Problems. Nuclear Engineering and Design, Vol. 4 p. 276, 1966.

3.	Campion-Renson, Anne and Crochet, M.J. - On the Stream Function - Vorticity Finite Element Solution of Navier Stokes Equations. Int. J. Num. Meth. Engng, Vol. 12 pp. 1809-1818, 1978.

4.	Ikegawa, M. - A New Finite Element Technique for the Analysis of Steady Viscous Flow Problems. Int. J. Num. Meth. Engng., Vol. 14 pp. 103-113 1979.

5.	Moult, A., Burley, D. and Rawson, H. - The Numerical Solution of Two-Dimensional Steady State Flow Problems by the Finite Element Method. Int. J. Num. Meth. Engng., Vol. 14 pp. 11-35. 1979.

6.	Olson, M.D. - Variational-Finite Element Method for Two Dimensional and Axisymetric Navier-Stokes Equations. Finite Element in Fluids, Vol. 1 pp 57-72, Ed. Gallagher, R.H., Oden, J.T., Taylor, C. and Zienkiewicz, O.C., John Wiley and Sons, 1978.

7.	Oden, J.T. - Finite Element Analogue for Navier-Stokes Equations. J. Engng. Mech Div., ASCE, 96(EM4) pp. 529-534, 1970.

8.	Kawahara, M. et al - Steady and Unsteady Finite Element Analysis of Incompressible Viscous Fluid. Int. J. Num. Meth. Engng., Vol. 10 pp. 437-456, 1976.

226

9. Olson, M.D. and Tuann, S.Y. - Primitive Variables versus Stream Function Finite Element Solutions of the Navier-Stokes Equations. Finite Element in Fluids, Ed. Gallagher, R.H. et al, John Wiley and Sons, 1978.

10. Temam, R. and Thomasset, F. - Solution of the Navier - Stokes Equations by Finite Element Method. Int Proc. 4th Int. Conf. Num. Meth. Fluid Mech., pp. 392-402, Ed. Richtmyer, R.D., 1974.

11. Malkus, D.S. - Penalty Methods in Finite Element Analysis of Fluids and Structures. Nuclear Engineering and Design, Vol. 57 pp. 441-448, 1980.

12. Hughes, T.J.R. et al - Finite Element Analysis of Incompressible Viscous Flows by the Penalty Function Formulation. J. Comp Physics, Vol. 30 pp. 1-60, 1979.

13. Hughes, T.J.R. - High Reynolds Number Steady Incompressible Flow by Finite Element Method. Finite Element in Fluids, vol. 3, Ed. Gallagher, R.H. et al, John Wiley and Sons, 1978.

14. Reddy, J.N. and Safake, A. - A Comparison of a Penalty Finite Element Model of Natural Convection in Enclosures. J. Heat Transfer, Vol. 102 pp. 659-666, Nov. 1980.

15. Heinrich, J.C. and Marshall, R.S. - Viscous Incompressible Flow by a Penalty Function Finite Element Method. Computer and Fluids, Vol. 9 pp 73-83, 1981.

16. Engleman, M.S. et al - Consistent Versus Reduced Integration Penalty Methods for Incompressible Media using several Old and New Elements. Int. J. Num. Meth. Fluids, Vol. 2 pp. 25-42, 1982.

17. Reddy, J.N. - On Penalty Function Method in the Finite Element Analysis of Flow Problems. Int. J. Num. Meth. Fluids, Vol. 2 pp. 151-171, 1982.

18. Hughes, T.J.R. - Generalisation of Selective Integration Procedures to Anisotropic and Non-linear Media. Int. J. Num. Meth. Engng., Vol. 15 pp. 1413-1418, 1979.

19. Malkus, D.S. and Hughes, T.J.R. - Mixed Finite Element Method - Reduced and Selective Integration Techniques a Unification of Concepts. Comp. Meth. Mech. Engng., Vol. 15 pp 63-81, 1978.

20. Hood, P. and Taylor, C. - Navier Stokes Equations using Mixed Interpolation in Finite Element Method in Flow Problems. University of Alabama, pp. 121-132. Ed. Oden, J.T., Huntsville Press, 1974.

21. Tuann, S.Y. and Olson, M.D. - A Study of Various Finite Element Solution Methods for the Navier Stokes Equations. Struct. Res. Series, Report 14, Dept. Civil Engng., University of British Colombia, June 1967.

22. Finlayson, B.A. - The Method of Weighted Residual and Variational Principles, Academic Press 1972.

23. Prassas, A.G. - Finite Element Analysis of Heat Transfer in Fluid and Solid Media, Ph.D Thesis, Imperial College, London University, to appear.

24. Kawahara, M., Yoshimura, N., Nakagawa, K. and Ohsaka, H., Steady Flow Analysis of Incompressible, Viscous Fluid by the Finite Element Method. Theory and Practice in Finite Element Structural Analysis. Ed. Yamada, Y., and Gallagher, R.H. University of Tokyo Press, pp. 557-572.

25. Argyris, J.H., and Mareczek, G., Finite Element Analysis of Slow, Incompressible Viscous Fluid Motion. Ing. Arch. 43, pp. 92-109, 1974.

26. Nickell, R.E., Tanner, R.I. and Caswell, B, The Solution of Viscous, Incompressible Jet and Free-Surface Flows using Finite Element Methods. J. Fluid Mech. 65, Part 1, pp. 189-206, 1974.

27. Gartling, D.K., and Becker, E.B., Finite Element Analysis of Viscous Incompressible Fluid Flow. Comp. Meth. Appl. Mech. Engng. 8 pp. 51-60, 8 pp127-138, 1976.

28. Gartling, D.K., et al - A Finite Element Convergence Study for Accelerating Flow Problems. Int. J. Num. Meth. Engng. Vol. 11, pp. 1155-1174, 1977.

29. Strada, M. and Heinrich, J.C., Heat Transfer in Natural Convection at High Rayleigh Numbers in Rectangular Enclosures: A Numerical Study. Num. Heat. Tranf. Vol. 5. pp. 81-93, 1982.

CHAPTER 9

TRANSIENT TWO-DIMENSIONAL PHASE CHANGE WITH CONVECTION, USING
DEFORMING FINITE ELEMENTS

Mary Remley Albert[1]
Kevin O'Neill[2]

[1] Mathematician, U.S. Army Cold Regions Research and Engineering Laboratory, Hanover, N.H.
[2] Research Civil Engineer, U.S. Army Cold Regions Research and Engineering Laboratory, Hanover, N.H.

1. INTRODUCTION

Moving mesh techniques have proven to be powerful for
handling moving boundary problems. They provide a means of
accurately defining the location of a moving boundary at each
time step, and in the case of a phase change situation, avoid
the obvious problem of the possible existence of two phases
within a single element. Moving mesh techniques are particularly warranted when flow in the unfrozen zone is considered.
Clean specifications of both fluid and thermal boundary conditions are possible when phase boundaries and element boundaries
always coincide.

In the past, the difficulty of specifying systematically
the location and movement of interior mesh has tended to dis-
courage implementation of these techniques (e.g. refs. 4, 12).
In this paper, recent advances in automatic mesh generation are
used to specify the interior mesh smoothly and simply at each
time step. Linear triangular elements are used, and at each
time step node locations are generated using a transfinite
mapping technique.

Two example problems are considered to demonstrate the
versatility of the method. The first is that of two-dimensional axisymmetric solidification phase change in a pipe with
turbulent flow, under the assumption that a specified head drop
drives the flow. As freezing progresses and the head drop
remains constant, the flow slows until either stopping or
achieving a nonzero steady state. A convective thermal boundary condition is imposed at the moving phase boundary. The
overall flow rate is determined from the head drop using a drag
coefficient dependent in part on the Reynolds number. Thus the

overall flow rate, the fluid-solid heat transfer, and the consequent water temperature all depend on the evolving shape of the phase interface at each cross section of flow. The second example problem addresses Cartesian two-dimensional solidification in a porous medium with natural convection. In this case latent heat balance and advancement of the solidification front are determined from temperature gradients calculated within each phase, and the front is considered to be a streamline. In both problems phase change is isothermal, and a smooth front evolves as the mesh deformation responds to the heat and fluid flow patterns.

2. FINITE ELEMENT FORMULATION

In what follows we present the numerical formulation for the axisymmetric case. The Cartesian case proceeds analogously (see ref. 13 for details). The heat conduction equation governs behavior in the frozen zone:

$$C \frac{\partial T}{\partial t} = \nabla \cdot (K \nabla T) \qquad (1)$$

Motion of the solid-liquid interface is determined by the temperature gradient in the frozen region and the heat flux from the flowing water.

$$L \frac{d\underset{\sim}{s}}{dt} = K \nabla T - \underset{\sim}{q} \qquad (2)$$

where ds/dt is the rate of normal motion of the boundary, L the latent heat of fusion per unit solid volume, and q heat flux impinging from the liquid.

The finite element method formulation for equation (1) on a moving and deforming mesh has been presented previously [10,14]. One-dimensional mesh boundary motion was determined from the physical boundary conditions, while interior mesh motion was calculated using simple expediences. Subsequently a pseudo-elastic mesh deformation system was developed to control interior mesh motion in two dimensions [9], which was then analyzed in the context of other moving mesh methods and applied to Cartesian two-dimensional phase change [8]. The simplicity and flexibility of the transfinite mapping system of mesh generation [6] recommended it for interior mesh specification, and it too was applied to Cartesian two-dimensional freezing without convection [2]. Here we apply the mapping technique to the axisymmetric case as well.

The general formulation for (1) is the same as equation (11) in Ref. 10, which translates to an axisymmetric formulation as

$$\frac{dT_j}{dt} \int_{\Omega} C\phi_i\phi_j r \ drdz - T_j \int_{\Omega} \left[C\phi_i \frac{d\underset{\sim}{x}}{dt} \cdot \nabla\phi_j - \nabla\phi_i \cdot K\nabla\phi_j \right] r \ dr \ dz$$

$$- T_j \int_{\Gamma} \phi_i \underset{\sim}{n} \cdot K\nabla\phi_j d\Gamma = 0 \qquad (3)$$

where $\Omega(t)$ is the solid phase spatial domain, $\Gamma(t)$ its boundary, and $\underset{\sim}{n}$ a unit normal to Γ. The T_j are nodal temperature values, and the ϕ_j are linear triangular basis functions in (r, z). Conceptually, integration has been performed over 2π with respect to θ, on ring elements, and thus r appears as shown. In this moving mesh system, the treatment of spatial derivatives proceeds essentially as usual, however time dependence of the mesh gives rise to the second term in (3).

The mesh velocity and r are expressed as

$$\frac{d\underset{\sim}{x}}{dt} = \frac{d\underset{\sim m}{x}}{dt} \ \phi_m \ ; \quad r = r_m\phi_m \qquad (4)$$

where $\underset{\sim m}{x}$ and r_m are nodal values of (r,z) and of r, respectively.

Thus all integrals in (3) may be evaluated with simple closed form expressions, assuming C and K are constant over each element. Mesh velocity varies through time and is continuous though arbitrarily heterogeneous in space. Incorporation of the mesh velocity in the governing equation through all space, together with time dependence of the mesh, effectively translates the problem into a moving and deforming coordinate system. This means that the T_j apply directly to moving nodes, and are not obtained from recurrent interpolation between fixed nodes.

In principle, the implicitness (time weighting) in the expression of the boundary condition (2), of the space derivatives, and of the mesh and mesh velocity may all be specified separately. Here no iterations were performed on the solution of (1), (2) within each time step. Given current solution values, a forward projection of the boundary and hence of the entire mesh was made at the beginning of each time step on the basis of (2), which was then not included simultaneously in the subsequent solution of (1). Space derivatives were evaluated at the three-quarter time level, with mesh configuration centered in time, and dx/dt evaluated by forward difference (see refs 10, 11 and 14 for more accurate systems).

Boundary mesh motion was calculated from (2) using a scheme which conserves energy integrally ([8], [9]), to the

extent that the energy calculation on the interface may be
performed using solution gradients only. It can be shown that
this introduces an error, which is small in this case, which
may be corrected by inclusion of heat capacity effects in a
more general finite element expression of the boundary heat
balance ([11]). Without this correction the expression used
here for motion of the jth boundary node is:

$$\frac{d\underset{\sim}{x}}{dt} \cdot \int_{\Gamma} \underset{\sim}{n}\phi_j d\Gamma = \frac{1}{L} \int_{\Gamma} \left(K \frac{dT}{\partial n} - q\right)\phi_j d\Gamma \tag{5}$$

(no summation over j).

The direction of the vector dx_j/dt may be chosen to be that
of the integral on the left hand side, or that of a boundary
line along which the node must travel. Given any chosen direction, equation (5) determines a consistent magnitude of the
node velocity.

3. USE OF TRANSFINITE MAPPINGS

Haber et al. [6] describe the transfinite mapping technique as a Boolean combination of projectors, where a projector
performs a linear operation to map a true surface onto an
approximate surface. Essentially, for a four-sided region, the
mapping represents the region in a curvilinear coordinate
system formed by mapping a unit square onto the region.
Consider a four-sided region where u and v are normalized
coordinates extending over two adjacent sides. If $\psi_1(u)$,
$\psi_2(u)$, $\xi_1(v)$, and $\xi_2(v)$ are the boundary curves and $X(0,0)$,
$X(1,0)$, $X(0,1)$, $X(1,1)$ are corner locations, then interior
locations are given by

$$X(u,v) = (1-v)\psi_1(u) + v\psi_2(u) + (1-u)\xi_1(v) + u\xi_2(v)$$

$$- (1-u)(1-v)X(0,0) - (1-u)vX(0,1) - uvX(1,1) \tag{6}$$

$$- u(1-v)X(1,0) \qquad 0 \le u \le 1, \, 0 \le v \le 1$$

A mesh generated by this system for arbitrary boundary curves
is shown in Figure 1.

In the axisymmetric problem considered here, one side of
the four-sided region is the pipe wall, one side is the location of the solid-fluid phase interface, and the other two
sides represent the upstream and downstream boundaries of the
problem. The edges of the region are defined using piecewise
straight lines between boundary nodal coordinates for the
finite element formulation. The mapping is used to dictate the
location of the interior nodes by selecting sets of lines in
(u,v) passing through those boundary nodes, and mapping their

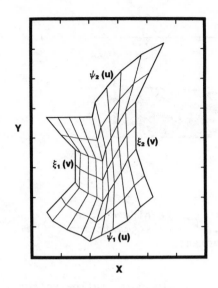

Figure 1. Example of finite element mesh generated from its boundary curves using transfinite mapping.

intersections to (r,z). Specifying interior node location in this manner ensures that the mesh reflects boundary curvature information for a small computational effort. This technique may be used for irregular boundaries, and thus represents a general approach to the problem. Refer to Haber et al. [6] for a more complete discussion and illustration of the technique. We note that slope discontinuities in the bounday curves are accommodated, and in general the interior mesh lines reflect the domain boundary in a well graded fashion.

4. TREATMENT OF THE AXISYMMETRIC CASE

Determination of the flow rate and boundary heat flux

For heat transfer to the boundary of a fully developed turbulent flow, Karlekar and Desmond [7] recommend the formula of Petvkhov and Popov to determine the heat transfer coefficient, h, from the Reynolds number Re and the friction factor f:

$$\frac{hD}{K} = \frac{(f/8) \text{ Re Pr}}{1.07 + 12.7 \ (f/8)^{1/2} \ (\text{Pr}^{2/3} -1)} \tag{7}$$

The friction factor is determined from the Filonenko equation:

$$f = (1.82 \ \log_{10}(\text{Re}) - 1.64)^{-2} \tag{8}$$

For any particular ice configuration, the flow rate through the pipe and heat transfer to the ice can be determined. To do so, one segments the pipe longitudinally into

234

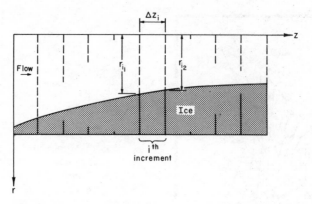

Figure 2. Cross section of the pipe on one
side of the centerline, showing segmentation
into computational increments.

increments, as shown in Figure 2. Assuming some inflow velocity and marching down the pipe, one applies conservation of
mass and equation (8) at each increment to obtain incremental
head drops. Summing these, one then iterates on the inflow
velocity to converge on the restriction that the head drop over
the total length of pipe is some specified value. Knowing the
average cross sectional velocity for each increment, one may
calculate the heat transfer coefficient for each segment of
interface (element boundary on the freezing front) using
equations (7) and (8). Given the inflow temperature and h for
each increment, an energy balance then determines the water
temperature along the pipe. Since the pipe temperature is
constant and cold over the entire length, the water temperature
decreases along the length. Finally, the heat flux into the
solid-liquid interface along each increment is calculated for
this time step from the local heat transfer coefficient and the
water temperature. This heat flux q is used in equation (5) to
specify the motion of the phase interface. Further details of
these calculations are given by Albert [1].

The total simulation fits together as follows. An initial
thin ice layer is assumed, with an initial temperature distribution. Heat fluxes are calculated for each increment of the
freezing front, the front is moved, a new mesh generated, and
new temperatures are calculated within the ice. This process
is repeated until the flow reaches a nonzero steady state, or
the pipe freezes shut. Strictly speaking, as the ice advances
the velocity may slow so much that equations (7) and (8) should
be replaced by equations suitable for laminar or transition
flows. However, this was not considered here. In addition, it
has been observed that certain conditions will produce a wavy
interface ([3],[5]). Here we assume that those conditions are
avoided.

Results

As a preliminary test, the model was run against the analytical solution for a radial freezing problem with good results ([1]). Then the model was run for two axisymmetric cases, one in which the ice thickness reaches steady state, and one in which the pipe freezes shut. In both cases, the pipe radius was 0.6 cm, the length 5.0 m, and the pipe temperature was −18°C. The phase change boundary was kept at 0°C, and zero flux boundary conditions were applied in the ice normal to the pipe ends. A uniform initial ice layer 0.05 cm was assumed, and the initial temperature profile was linear between the

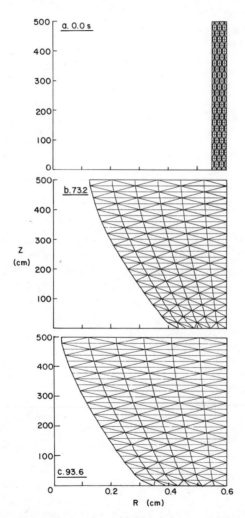

Figure 3. Mesh and ice configuration in the pipe a) initially,
b) at an intermediate time, and
c) just before freeze up.

freezing front and the pipe. The initial mesh is illustrated in Figure 3a. Fluid properties for water were assumed as follows:

$$K = 1.34 \times 10^{-3} \text{ cal/cm sec } °C$$
$$C = 1.0 \text{ cal/cm}^3 °C \tag{9a}$$

with the kinematic viscosity

$$\nu = 1.79 \times 10^{-2} \text{ cm}^2/\text{sec} \tag{9b}$$

with ice characteristics

$$K = 5.28 \times 10^{-3} \text{ cal/cm sec } °C$$
$$C = 0.4459 \text{ cal/cm}^3 °C \tag{9c}$$
$$L = 80 \text{ cal/cm}^3$$

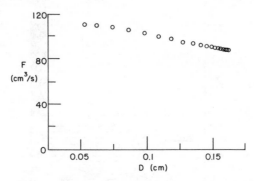

Figure 4. Volumetric flow rate (F) through the pipe as a function of frozen thickness at the outflow (D).

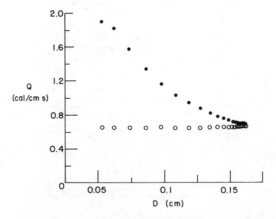

Figure 5. As frozen thickness at the outflow increases, heat flow at the interface through the ice (solid dots) and from the liquid (open circles) come to equilibrium, case 1.

In case 1 the fluid inlet temperature was maintained at
30°C and the head drop along the pipe was 400 cm. Figure 4
illustrates the volumetric flow rate, F, as a function of
frozen thickness; ultimately the flow rate reaches a steady
value. Figure 5 shows that the heat flux to the interface from
the fluid and the heat flux through the ice came to a balance
as the interface approached its final position, shown in Figure
6.

For the second axisymmetric case, the inlet fluid temperature was maintained at 60°C, and the head drop was 30 cm. Here
the pipe froze shut; one sees from Figure 7 that the flow rate
decreased to zero. The flux from the fluid at the cold end of
the pipe did not come to a balance with that from the frozen
layer (Fig. 8). Figure 3c shows the mesh shortly before
freeze-up.

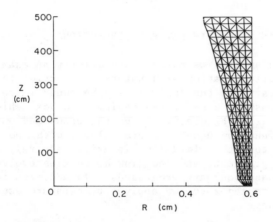

Figure 6. Steady state ice configuration in the pipe, case 1.

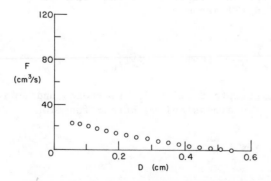

Figure 7. Flow rate through the pipe
as a function of frozen thickness at
the outflow, case 2.

238

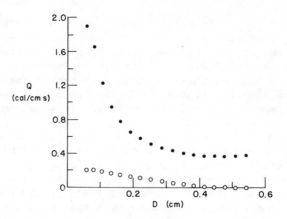

Figure 8. Comparable to Figure 5, but for case 2, indicating freeze up.

6. CARTESIAN CASE WITH NATURAL CONVECTION

In this section we pursue very briefly an example problem featuring solidification with natural convection in a porous medium. Details of the analysis may be found in ref. 13. In short, we consider a saturated medium in a box, which has been discretized as in Figure 9. Standard porous medium temperature and stream function equations are solved, with the distinguishing feature that the fluid flow is driven entirely by the temperature gradient, and the flow boundaries evolve continually as the phase boundary progresses. In addition to equation (1) for the frozen zone, we apply a temperature equation in the unfrozen zone

$$C \frac{\partial T}{\partial T} + c_r \underset{\sim}{v} \cdot \nabla T - \nabla \cdot (K \nabla T) = 0 \qquad (10)$$

where c_r is the heat capacity of the fluid alone. The interface condition (2) becomes

$$L \frac{d \underset{\sim}{s}}{dt} = (K \nabla T)_f - (K \nabla T)_u \qquad (11)$$

where the subscripts f and u denote frozen and unfrozen, respectively. The dimensionless stream function is determined by

$$\nabla^2 \psi = \frac{Ra}{\theta_w} \frac{\partial \theta}{\partial x} \qquad (12)$$

where Ra is the Rayleigh number, θ is a dimensionless temperature, and θ_w refers to the warm side boundary. Special measures are taken to achieve $C°$ continuous velocities from the

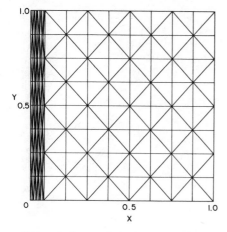

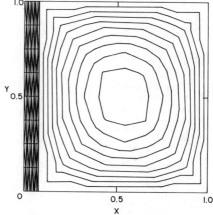

Figure 9. Mesh for natural convection case, shortly after start up. Dense area at left is frozen zone, the rest is unfrozen.

Figure 10. Streamlines and frozen zone shape shortly after start up in natural convection case.

solution of (12) over the linear elements. Details of that process, the scaling system, parameter values, and the numerical formulation as a whole are given in ref. 13. We note that, in addition to the mesh convection, the discretized equations will also contain a significant physical convection term in the unfrozen zone. This term couples the flow calculation to the temperature determination; the right hand side of (12) couples the temperature solution to the flow determination; and both together produce values of the temperature gradients in (11). In part because equation (11) determines evolution of the overall phase configuration and flow boundaries, it accomplishes coupling of all parts of the computation with all other parts.

The system is driven by the boundary conditions: The top and bottom of the box were insulated, while the left and right sides were kept equally below and above freezing respectively, such that a Rayleigh number of 100 resulted initially. A linear density fluid was assumed, and the streamline pattern in Figure 10 results shortly after startup. As solidification proceeds, the circulation loses symmetry, passing through the configuration in Figure 11 and achieving the steady state shown in Figure 12. Plots of the isotherms at the same points in time (Figs. 13-15) reveal distortions wrought by the convection, relative to profiles expected from pure conduction. One sees that convection intensifies as freezing proceeds. This is because the advancement of the isothermal freezing front towards the constant temperature warm side induces larger lateral temperature (density) gradients.

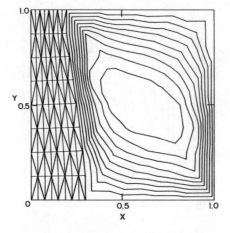

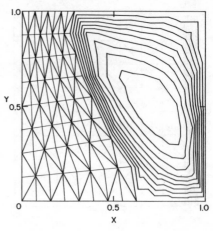

Figure 11. Streamline pattern and phase configuration at intermediate time in natural convection case.

Figure 12. Steady state streamline and phase configuration.

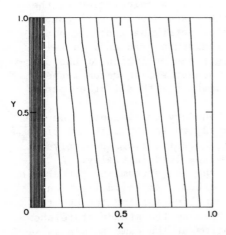

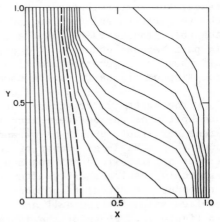

Figure 13. Isotherm pattern and frozen zone shape shortly after start up in natural convection case.

Figure 14. Isotherms at same intermediate time as for Figure 11.

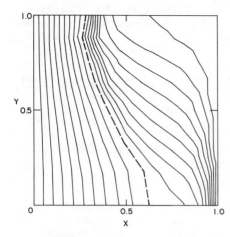

Figure 15. Steady state
isotherm pattern.

7. SUMMARY AND COMMENTS

The use of transfinite mappings in a moving mesh scheme
provided an easy, efficient and accurate way to model these
moving boundary problems. A clear definition of the phase
change front is obtained, so that heat and fluid flow may be
handled separately in each phase, and the two successfully
joined at the boundary. The problems considered here result in
fairly uniform and well behaved meshes, and do not test the
limit of the utility of the transfinite mapping technique.
Related work underway using an elastic grid system treats more
irregular regions, coupled solute transport, and more complex
flows [15]. Efficiency comparison tests with a transfinite
mapping formulation will be run under these more complex
conditions.

8. ACKNOWLEDGMENTS

The authors wish to express their gratitude to Prof.
Daniel R. Lynch for help in identifying this problem, and to
Dr. George D. Ashton for help in obtaining heat transfer and
drag coefficients.

9. REFERENCES

1. ALBERT, M.R. - Modeling Two-Dimensional Freezing Using
   Transfinite Mappings and a Moving Mesh Finite Element
   Technique, Masters Thesis, Dartmouth College, 1983.

2. ALBERT, M.R. and K. O'NEILL - Adaptive Meshes by Transfinite Mappings for Phase Change Problems, Adaptive
   Computational Methods for Partial Differential Equations,
   Babuska et al. (eds.), SIAM, Philadelphia, pp. 85-110,
   1983.

3. ASHTON, G.D. and J.F. KENNEDY — Ripples on the Underside of River Ice Covers, Proc. ASCE J. Hydraul. Div., Vol. 98, No HY9, pp. 1603-1624, 1972.

4. GARTLING, D.K. — Finite Element Analysis of Convective Heat Transfer Problems with Change of Phase, in Numerical Methods in Laminar and Turbulent Flow, C. Taylor, K. Morgan, and C.A. Brebbia, Eds., Pentech Press, London, 1978.

5. GILPIN, R.R. — Ice Formation in a Pipe Containing Flows in the Transition and Turbulent Regimes, Trans. ASME J. Heat Transf., Vol. 103, pp. 363-368, 1981.

6. HABER, R., M.S. SHEPHARD, J.F. ABEL, R.H. GALLAGHER, and D.P. GREENBERG — A General Two-Dimensional Graphical Finite Element Preprocessor Utilizing Discrete Transfinite Mappings. Int. J. Num. Meth. Eng., Vol. 17, pp. 1015-1044, 1981.

7. KARLEKAR, B.V. and R.M. DESMOND — Engineering Heat Transfer, West Publishing Co., p. 350, 1977.

8. LYNCH, D.R. — A Unified Approach to Simulation on Deforming Elements with Application to Phase Change Problems. J. Comp. Phys., pp. 387-411, 1982.

9. LYNCH, D.R. and K. O'NEILL — Elastic Grid Deformation for Moving Boundary Problems in Two Space Dimensions, Third Intl. Conf. Finite Elements Wat. Resour., S.Y. Wang et al. (Eds.), pp. 3.67-3.76, 1980.

10. LYNCH, D.R. and K. O'NEILL — Continuously Deforming Finite Elements for Solution of Parabolic Problems With and Without Phase Change. Int. J. Num. Meth. Eng., Vol. 17, pp. 81-96, 1981.

11. LYNCH, D.R. and J.M. SULLIVAN — Heat conservation in moving mesh finite element calculations, J. Comput. Phys., in press.

12. MORGAN, K. — A Numerical Analysis of Freezing and Melting with Convection. Comp. Meth. Appl. Mech. Eng., Vol. 28, pp. 275-284, 1981.

13. O'NEILL, K. and M.R. ALBERT — Computation of Porous Media Natural Convection Flow and Phase Change, 5th Int. Conf. Finite Elements Wat. Resour., June, 1984.

14. O'NEILL, K. and D.R. LYNCH — A Finite Element Solution for Freezing Problems, Using a Continuously Deforming Coordinate System, Chapter 11 in Num Meth Heat Transfer, R.W.

Lewis, K. Morgan, and O.C. Zienkiewicz (Eds.), Wiley Interscience, 1981.

15. SULLIVAN, J.M., D.R. LYNCH, and K. O'NEILL - Finite Element Solution of Ice Crystal Growth in Subcooled Sodium-Chloride Solutions, Intl. Conf. Adv. Num. Meth. Eng: Theory and Appl., 1985, to appear.

CHAPTER 10

APPLICATIONS OF CONTROL VOLUME ENTHALPY METHODS IN THE
SOLUTION OF STEFAN PROBLEMS.

V. R. VOLLER and M. CROSS
School of Mathematics, Statistics and Computing, Thames
Polytechnic, London S.E.18. 6PF

1. INTRODUCTION

The modelling of many physical problems in heat and mass
transfer requires the analysis of a phase change phenomenon.
Examples include the solidification of steel [1] and the dif-
fussion of oxygen in the lung [2]. A range of techniques,
both analytical and numerical, for dealing with phase change
problems have been developed over the years. An extensive bi-
bliography has been prepared by Tarzia [3]. Summaries of existing analytical methods may be found in Carslaw and
Jaeger [4] and Crank [5]. Numerical techniques are by far the
most common means for analysis of phase change problems and
most of the available approaches are well documented in the
reviews of Fox [6], Crank [7, 8] and Furzeland [9].

An integral part in the solution of a phase change problem is tracking the moving boundary on which the phase change
occurs. Standard numerical approaches based on a fixed grid
prove difficult because the resulting solution procedure needs
to account for this continuously moving boundary over a discrete domain. As a result many of the phase change numerical
techniques which have been developed involve a grid which de-
forms as the phase boundary moves. The relative simplicity
of a fixed grid solution may be retained by writing the governing equations in conservation form via the introduction of an
enthalpy function. In this way, with a correctly defined enthalpy, the full effects of a phase change may be modelled
even though the exact position of the phase change boundary
is unknown.

The classical phase change problem is that of melting/
freezing via the conduction of heat, i.e. the so called Stefan problem. In this chapter a number of straight forward
techniques, based on enthalpy formulations, for the numerical

solution of phase change problems are presented and examined. The majority of the numerical solutions discussed will be based on control volume discretizations [10]. The basic ideas presented, however, may have applications in alternative numerical techniques, in particular finite element methods.

2. THE STEFAN PROBLEM

The governing equations of the Stefan melting/freezing problem may be derived on considering a region R of pure material separated into solid, R_s, and liquid, R_ℓ, sub-regions by a sharply defined interface Σ. Then, if conduction is the only mechanism of heat transfer and there is no density change across the interface, two Fourier heat conduction equations may be written, one for the solid region

$$\rho_s \, C_s \, \partial T/\partial t = \text{div} \, (K_s \, \text{grad} \, T) \tag{1a}$$

and one for the liquid region

$$\rho_\ell \, C_\ell \, \partial T/\partial t = \text{div} \, (K_\ell \, \text{grad} \, T) \tag{1b}$$

The interface Σ is continuously moving through the region, its direction dependent on the nature of the phase change, and on forming the heat balance across the interface,

$$\left(K\frac{\partial T}{\partial n}\right)_s - \left(K\frac{\partial T}{\partial n}\right)_\ell = \rho_m \, \lambda \, U \tag{2}$$

where

$$\lambda = \begin{cases} L \text{ for freezing} \\ -L \text{ for melting} \end{cases}$$

is the latent heat associated with the phase change, n is the distance along the local normal to Σ and U is the velocity of Σ along this normal. Equations (1) may be referred to as the Stefan formulation. Equation (2) which states that the net flow of heat across the phase change interface equals the amount of heat used in undergoing the phase change, is often called the "Stefan" condition.

3. THE ENTHALPY FORMULATION

By defining an enthalpy function $H(T)$ to be the sum of latent and sensible heats [1] equations (1) and (2) can be reduced to a single equation [11]

$$\frac{\partial H(T)}{\partial t} = \text{div (K grad T)} \tag{3}$$

i.e. the enthalpy formulation. On considering the energy conservation for a volume V in the absence of external work and
internal energy sources, an integral form of equation (3) may
be written [12].

$$\frac{d}{dt} \int_V \rho H \, dV = \int_A K \text{ grad T } \underline{\hat{n}} \, dA \tag{4}$$

The relationship between the enthalpy and temperature is
often expressed as (see Figure 1a):

$$H(T) = \begin{cases} CT & T < T_m \\ CT + L & T > T_m \end{cases} \tag{5}$$

From a practical viewpoint, however, the jump discontinuity
in this relationship is not desirable. There are three approaches to overcome this problem. The function H(T) can be
assumed or measured experimentally over the region of interest
(see Eyres, [11] Albasiny, [13] Lockwood [14] and Longworth
[15]). The jump in H(T) at $T = T_m$ can be smoothed over a
small interval, Szekely and Themelis, [1] Szekely and Lee,
[16] Meyer, [17] and Bonacina, et al [18] (see Figure 1b)
This step may result in a function of the form

$$H(T) = \phi^{-1}(T) = \begin{cases} CT & T_m - \varepsilon \geq T \\ CT + L(T - T_m + \varepsilon)/2\varepsilon & T_m - \varepsilon < T < T_m + \varepsilon \\ CT + L & T \geq T_m + \varepsilon \end{cases} \tag{6}$$

The parameter ε is a small temperature interval over which
the phase change is assumed to occur. In physical problems
a phase change temperature range is common. The last of the
three methods of circumventing the discontinuous nature of
H(T) is to express T as a function of H, Rose [19],
Atthey [20, 21].

248

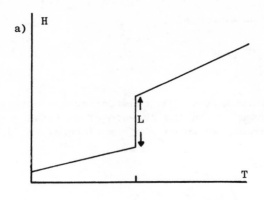

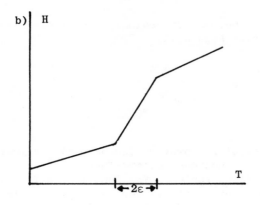

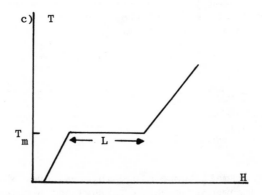

Figure 1   Enthalpy Temperature Relationships

Then

$$
T(H) = \phi(H) = \begin{cases} H/C & , \ C(T_m-\epsilon) > H \\[2ex] T_m + \epsilon\dfrac{(H-CT_m- L/2)}{C\epsilon + L/2} & , \ C(T_m-\epsilon) \geq H \geq C(T_m+\epsilon)+L \\[2ex] (H-L)/C & , \ H < C(T_m+\epsilon)+L \end{cases} \tag{7}
$$

which is a piecewise continuous function for all values of $\epsilon \geqslant 0$, in particular $\epsilon = 0$ (see Figure 1c).

First use of the enthalpy formulation for the solution of phase change problems in heat transfer are reported by Dusinberre [22] in 1945 and Eyres et al [11] in 1946. Early applications include: Price and Slack [23] who developed an enthalpy solution for freezing in the semi infinite plane; Albasiny [13] who considered solidification problems in finite slabs; and Baxter [24] who investigated the freezing of cylinders. Numerical analysis of enthalpy formulation solutions was first reported by Kamenomostskaja [25] and Oleinik [26]. More recent work in this area is summarised in Elliott and Ockendon [27]. Equivalence between the enthalpy formulation and the Stefan formulation has been demonstrated by Solomon [28] and Shamsundar and Sparrow [12].

4. NUMERICAL SOLUTIONS

4.1 Control Volume Techniques

The first step in any numerical procedure is to cover the domain of interest by a discrete number of node points. The governing equations are then written in a discrete form in terms of values at node points and a numerical solution algorithm generated. In the numerical solution of equations which model flow phenomena a popular method is to associate with each node point a control volume. On approximating conservation of flows over these control volumes in terms of gradients between node points discrete equations result which form the basis of a numerical solution. Such techniques were first suggested by Dusinberre [22] and have recently been updated and fully expanded on by Patankar [10].

To illustrate the control volume technique consider a one dimensional heat conduction problem; typical nodes and control volumes are shown in Figure 2.

The flow of heat per unit area across the west face (W) of the ith control volume at time $t = j\delta t$ is

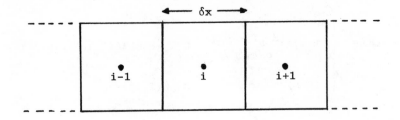

Figure 2.   Control Volumes

$$\left[K\frac{\partial T}{\partial x} \right]_{w,j\delta t}$$

which may be approximated as

$$K_w\, (T_{i,j} - T_{i-1,j})/\delta x$$

where $(*)_{i,j}$ indicates the value on the ith node point at time $j\delta t$. In the same manner the flow of heat per unit area across the east face (E) is

$$K_E\, (T_{i+1,j} - T_{i,j})/\delta x$$

Taking the ith node point as representative the rate of change of the enthalpy (i.e. the total heat content) of the ith control volume may be written as

$$\left(\rho\, \frac{dH_i}{dt} \right) \delta x$$

On assuming that the flow conditions prevail over the entire time interval $j\delta t$ to $j + 1\, \delta t$ conservation in the ith control volume gives

$$H_{i,j+1} = H_{i,j} + \frac{\delta t}{\rho \delta x^2} \left(K_E\, (T_{i+1,j} - T_{i,j}) - K_w (T_{i,j} - T_{i-1,j}) \right)$$

Further assuming that the conductivity is a constant the above scheme reduces to the familiar finite difference form

$$H_{i,j+1} = H_{i,j} + \frac{K\delta t}{\rho \delta x^2} (T_{i+1,j} - 2T_{i,j} + T_{i-1,j}) \qquad (8)$$

An identical finite difference scheme would be obtained if standard finite difference methods are used [29]. Equivalence between the control volume and standard finite difference techniques is often the case for simple problems. The subtle but important difference in using a control volume approach is that the physical characteristics of the flow system will always be accounted for.

4.2 Implicit Techniques

When used in conjunction with discrete forms of equations (6) or (7) the explicit scheme, equation (8), is relatively simple to use. There are two drawbacks in using an explicit approach. From physical considerations the assumption that during one time step the flow conditions at the start prevail throughout, may not be valid. In addition, from numerical considerations there are restrictions on the step sizes, δt, δx, etc. for stability.

To overcome the above problems a weighted control volume scheme may be derived, which in vector form is

$$\underline{H}^{j+1} = \underline{H}^j + \delta t(1-\theta) \underline{F}(\underline{T}^j) + \delta t\theta \underline{F}(\underline{T}^{j+1}) \qquad (9)$$

where $*^j$ is the vector of nodal values of $(*)$ at time $t = j\delta t$. The parameter θ in equation (9) may take values between 0 and 1. A value of $\theta = 0$ will give the explicit scheme, equation (8), with $\theta = \frac{1}{2}$ a "Crank-Nicolson" type scheme is derived. When $\theta = 1$ a fully implicit scheme results. Any scheme will be unconditionally stable for values of $\theta \geqslant \frac{1}{2}$. The vector $\underline{F}$ in equation (9) expresses the net heat flow into a control volume at the specified time step. The form that $\underline{F}$ takes depends on the problem in question. The following examples for the ith component of $\underline{F}$ are given.

a) A one dimensional cartesian problem with constant thermal properties

$$F_i = \frac{K}{\rho \delta x^2} (T_{i+1} - 2T_i + T_{i-1}) \qquad (10)$$

b) A one dimensional cartesian freezing problem with a

step change in the conductivity at the phase change interface

$$F_i = \frac{1}{\rho \delta x^2} [K_1(T_{i-1} - T_i) + K_2(T_{i+1} - T_i)] \qquad (11)$$

where

$$K_1 = K_2 = K_s \qquad \text{if} \qquad T_i < T_m$$

$$K_1 = K_s \quad K_2 = K_\ell \qquad \text{if} \qquad T_i = T_m$$

$$K_1 = K_2 = K_\ell \qquad \text{if} \qquad T_i > T_m$$

c) A cylindrically symmetric problem with constant thermal properties

$$F_i = \frac{K}{\rho \delta r^2} [(1+\tfrac{1}{2}i)T_{i+1} - 2T_i + (1-\tfrac{1}{2}i)T_{i-1}] \qquad (12)$$

d) A two dimensional problem in a square region (nhxnh) with node numbering from bottom left to top right and with $\delta x = \delta y = h$

$$F_i = \frac{K}{\rho h^2} (T_{i-1} + T_{i-n} - 4T_i + T_{i+1} T_{i+n}) \qquad (13)$$

Note that the forms of $\underline{F}$ given in equations (9)-(13) are for internal control volumes. The components of $\underline{F}$ for control volumes adjacent to the domain boundary will need to take a different form to account for the boundary conditions.

The basic difficulty in using an implicit form of equation (9) is that a non-linear system of equations requires solution on each time step. Meyer [17] solves the non-linear system by writing the equations in terms of T alone via use of the function $\phi^{-1}(T)$, equation (6). Then a "Gauss-Seidel non-linear iterative scheme is employed for calculating the unknowns, viz T^{j+1}. The two drawbacks in this approach are:- 1) the form of the iterative scheme depends on the current estimate of T^{j+1} and 2) due to the jump discontinuity in $\phi^{-1}(T)$ when $\varepsilon = 0$ (see Figure 1a) the iterative scheme does not converge if ε is too small.

The problem of non convergence when ε is small may be overcome on writing equation (9) in terms of H alone via use of the function $\phi(H)$, which is piecewise continuous for all values of $\varepsilon \geqslant 0$. Shamsundar and Sparrow [12] and White [30, 31] propose this approach using an iterative scheme to solve the resulting non-linear equations in H^{j+1}. Longworth [15] also employs the function $\phi(H)$ but uses a Newton method in solving for H^{j+1}. In all these methods the problem of the solution scheme depending on the current iterative values is not eliminated.

An alternative method for soliving implicit forms of equation (9) has recently been proposed, Voller [32]. Equation (3) may be written as

$$\rho \frac{\partial}{\partial t} (CT + \Delta H) = \nabla.(K\nabla T) \qquad (14)$$

where ΔH is the latent heat component of the enthalpy given by

$$\Delta H = \begin{cases} 0 & T_m - \varepsilon \geq T \\ \dfrac{L}{2\varepsilon}(T-T_m + \varepsilon) & T_m - \varepsilon < T \leq T_m + \varepsilon \\ L & T > T_m + \varepsilon \end{cases} \qquad (15)$$

On re-arrangement equation (14) becomes

$$\rho C \frac{\partial T}{\partial t} = \nabla.(K\nabla T) + S \qquad (16)$$

where S is a latent heat source given by

$$S = -\rho \frac{\partial \Delta H}{\partial t} \qquad (17)$$

By defining $\underline{\Delta H}$ to be the vector of nodal latent heats a general finite difference scheme for equation (16) is

$$\underline{T}^{j+1} = \underline{T}^j + \frac{\delta t(1-\theta)}{C} \underline{F}(\underline{T}^j) + \frac{\delta t\theta}{C} \underline{F}(\underline{T}^{j+1}) + \underline{S}^{j+1} \qquad (18)$$

where

$$\underline{S}^{j+1} = \frac{\underline{\Delta H}^j - \underline{\Delta H}^{j+1}}{C}$$

When the value of S^{j+1} is known, guessed or otherwise equation (18) becomes a linear system which may be solved by standard techniques. Here the nodal enthalpies at each stage are simply given by

$$H_i^{j+1} = CT_i^{j+1} + \Delta H_i^{j+1} \tag{19}$$

A possible solution method for equation (18) is as follows. An initial guess for the nodal latent heat vector $\underline{\Delta H}_o^{j+1}$ is made. A suitable guess would be $\underline{\Delta H}_o^{j+1} = \underline{\Delta H}^j$ i.e. a zero latent heat phase change is assumed. Initial values for the components in $\underline{S}_o^{j+1}$ can then be calculated and the system of linear equations derived from equation (18) solved completely to give $\underline{T}_o^{j+1}$. From these values the latent heat source vector is updated and the system of equations solved for $\underline{T}_1^{j+1}$. This iterative cycle continues to convergence.

The methodology for updating the latent heat source vector , i.e. essentially updating ΔH^{j+1}, at the end of each iterative sweep is as follows. First the control volumes in which the phase change is occurring are identified on noting the control volumes in which

$$[T_i]_k^{j+1} < \varepsilon \qquad \text{and} \qquad [\Delta H_i]_k^{j+1} > 0$$

for a freezing problem, or

$$[T_i]_k^{j+1} > -\varepsilon \qquad \text{and} \qquad [\Delta H_i]_k^{j+1} < L$$

for a melting problem. In these control volumes the nodal latent heats are updated as

$$[\Delta H_i]_{k+1}^{j+1} = [\Delta H_i]_k^{j+1} + C([T_i]_k^{j+1} - T_i^*) \tag{20}$$

In the remaining control volumes the nodal latent heat is set equal to ΔH_i^j.

$$T_i^* = T_m + 2\varepsilon\left[[\Delta H_i]_k^{j+1} - L/2 \quad /L\right] \tag{21}$$

is the value that would be predicted for the nodal temperature T_i^{j+1} by using the current value of ΔH_i^{j+1} in equation (7).

In practice when updating $\underline{\Delta H}_i^{j+1}$ via equation (20) it should be checked that $[\Delta H_i]_{k+1}^{j+1} \varepsilon [0, L]$. When this is not the case it indicates that the phase change has been completed in the time interval $[j\delta t, (j+1)\delta t]$. Account of this may be taken on setting

$$[\Delta H_i]_{k+1}^{j+1} = 0 \quad , \qquad \text{Freezing}$$

or

$$[\Delta H_i]_{k+1}^{j+1} = L \quad , \qquad \text{melting}$$

The basic steps in the above implicit scheme are presented in flow sheet form in Figure 3.

4.3   Other Solution Techniques

Control volume techniques are clearly not the only means for the numerical solution of the enthalpy formulation. A great deal of work has been published outlining alternative enthalpy solutions. Some of these techniques are briefly discussed below.

Wood et al [33] develop an odd even hopscotch [34] finite difference technique for solution of the enthalpy formulation. The advantage of this approach is that an "explicit" type stable scheme is retained without severe restrictions on the step sizes δx and δt.

Due to the conservation form of the enthalpy formulation, equation (4), it is well suited to finite element techniques. An early finite element enthalpy technique is reported by Comini et al [35]. This technique has been updated by Morgan et al [36]. A drawback of these techniques is that it is necessary to introduce a phase change temperature range ε Therefore pure materials with single temperature phase changes cannot be accurately analysed. The reason for the introduction of the temperature range is that fixed, finite element grids are used. Single temperature phase change may be coped with by using front tracking methods combined with a deforming finite element grid. A finite element technique based on an enthalpy function that uses front tracking has been proposed by Li [37]. Recently, however, Rolph and Bathe [38] and

256

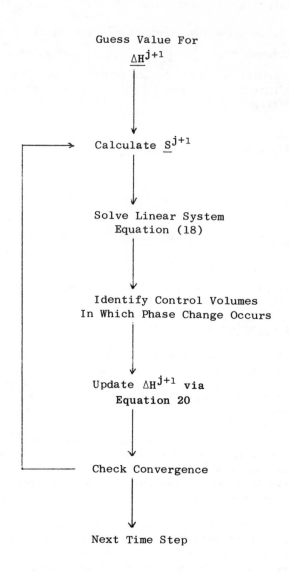

Figure 3.   Implicit Schme Outline

Winzell [39] have presented fixed grid finite element enthalpy methods which can cope with single temperature phase change. The basic feature in the work of Rolph and Bathe is the introduction of a source term which keeps track of the latent heat content of the finite elements within the solution domain. This approach bears a close resemblance to the implicit control volume technique outlined above. In fact control volume techniques and finite element techniques are often closely related. This is illustrated in the work of Ronel and Baliga [40] who develop a "control volume finite element technique" for solution of phase change problems based on an enthalpy formulation.

5. INTERPRETATION OF THE ENTHALPY

In the work that follows a number of control volume enthalpy schemes for solution of Stefan type problems will be presented. Many of these methods are built around an interpretation of enthalpy in a discretized region proposed by Voller [41]. Using the integral enthalpy formulation, equation (4), to express the conservation of heat in a control volume (i) in a solution domain undergoing a single temperature phase change, it may be shown that

$$\frac{dH_i}{dt} = \begin{cases} - L \dfrac{dS}{dt} & , \quad \text{Freezing} \\[2em] L \dfrac{dS}{dt} & , \quad \text{melting} \end{cases} \tag{22}$$

where S is the portion of the control volume which has undergone the phase change. Equation (22) states that while the solid/liquid boundary is in control volume .i the rate of change of the nodal enthalpy is proportional to the rate at which the control volume changes state. Since the nodal enthalpy H_i lies within the range

$$CT_m \leq H_i \leq CT_m + L$$

while control volume i changes state solution of equation (12) yields a relationship of the form

$$S = \begin{cases} (CT_m + L - H_i)/L & , \quad \text{freezing} \\[2em] (H_i - CT_m)/L & , \quad \text{melting} \end{cases} \tag{23}$$

Hence the amount of change of state that has occurred in the control volume is related to the nodal enthalpy value. An alternative form for equation (23) is

$$S = \begin{cases} (L - \Delta H_i)/L & , \quad \text{freezing} \\ \\ \Delta H_i/L & , \quad \text{melting} \end{cases} \qquad (24)$$

where ΔH_i is the latent heat content of the control volume as defined in equation (15). Equation (24) is a more general expression which is valid for phase changes that take place over a range.

In a one dimensional single temperature phase change equation (24) can be used directly to position the phase front from the numerically predicted $\underline{\Delta H}$ [42]. For multi dimensional problems the total extent of the phase change may be calculated from equation (24). In two dimensions, for example, if an area $A°$ is undergoing a phase change then the area which has changed state at any point in time is

$$A = S^* A° \qquad (25)$$

The parameter S^* is estimated from equation (24) by the following expression

$$S^* = \sum_{N}^{N*} S \qquad (26)$$

where N is the number of control volumes in $A°$ and N^* is the number of control volumes which have changed or are changing state.

6. ONE DIMENSIONAL APPLICATIONS

6.1 Direct Applications

In one dimension a standard test problem is that of freezing in the half plane. A material which may exist in a solid and liquid phase fills the half plane $x \geqslant 0$. For time $t \leqslant 0$ the material is in the liquid phase at temperature $T_I > T_m$ At time $t = 0$ the temperature at $x = 0$ is lowered and fixed at $T_o < T_m$. Therefore as time increases a layer of solid material will advance into the liquid. Conditions for a test problem of this type are given in Table 1. The reason for this choice of test problem is that the analytical solution is well known [4, 5] and $T_I > T_o$.

TABLE 1.   TEST PROBLEM 1

$$T_I = 2 \qquad T_o = -10 \qquad T_m = 0$$

$$K = 2 \qquad C = 2.5\ 10^6 \qquad L = 10^9$$

When the above test problem is solved using the explicit enthalpy method, equation (8), in conjunction with equation (7) the numerical results produce oscillations [18, 43].

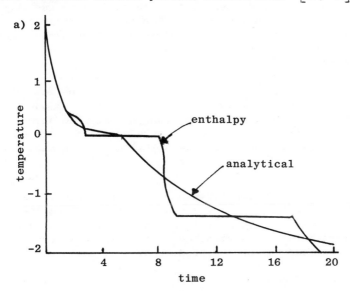

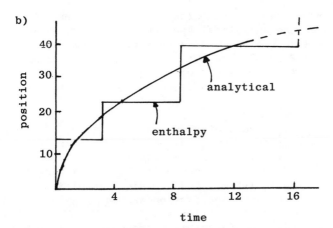

Figure 4. Comparison of analytic and direct enthalpy solutions in one-dimension: $[\delta x = .125,\ \delta t = 3600]$.

a) temperature history at x = .25
b) phase change movement

These oscillations are illustrated in Figure 4, by a comparison of enthalpy predictions and analytical solutions for the test problem in Table 1.

The reason for the step like predictions of the phase change movement are due to the fact that the phase change occurs in a control volume over a finite period of time. While the phase change occurs in the ith control volume the nodal enthalpy $H_i \in [0, L]$ and therefore, by equation (7), $T_i = 0$. Hence the phase front position, i.e. the zero isotherm, will remain fixed at $x = i\delta x$ while the ith control volume changes state. As the ith control volume completes its phase change the (i+1)th control volume will begin to change state and the numerical phase front will "jump" to position $x = (i+1) \delta x$.

The reasons for the series of plateaux in the temperature history numerical results follow directly from the manner in which the numerical phase front moves. The numerical phase front is fixed at $x = i\delta x$ for a number of time steps. Therefore as far as the numerical solution is concerned in the solid region, $x < i\delta x$, the problem solved is that of a region between two surfaces at fixed temperatures T_o and T_m. The steady state solution of this problem is rapidly reached and the plateaux formed. These remain until the numerical phase front jumps to its next location at which point a new steady state solution is attained. An analytical treatment of this behaviour has been presented by Bell [44].

6.2. Remedial Schemes

A feature of the direct enthalpy method is that there are points at which the analytical and numerical predictions agree. Below some remedial enthalpy schemes to minimise the oscillations in Figure 4 are outlined. The aim of such schemes is to predict these points of agreement and recover the enthalpy solution. Three classes of schemes will be presented.

Scheme 1: Latent heat at node. A close study of the results from a direct enthalpy solution of a cartesian problem reveals that when a latent heat contribution to a nodal enthalpy, H_i, is $L/2$, (i.e. when half the latent heat has been evolved), the analytical and numerical results are in close agreement. Such an observation was made by Price and Slack [23]. More recently Voller and Cross [42, 45, 46] have developed numerical procedures based around this observation. A theoretical basis is provided from equation (24), from which it may be seen that if $\Delta H_i = L/2$, S, the portion of the phase change completed in control volume i is 1/2. This means that in a one dimensional cartesian problem when

$\Delta H_i = L/2$ the phase front will pass through the ith node point. Therefore in a numerical enthalpy solution if at successive time steps H_i^{j+1} and H_i^j bound the value $CT_m + L/2$ the phase front can be assumed to have passed through the point $x = i\delta x$ during the time interval $[j\delta t, (j+1)\,\delta t]$. On assuming that the enthalpy change is linear in this time interval the time at which the phase boundary is at the point $i\delta x$ may be found from

$$t_i = (j + \chi)\delta t \tag{27}$$

where $\chi < 1$ is estimated via linear interpolation, in time, viz

$$\chi = \frac{(L/2 + CT_m - H_i^j)}{(H_i^{j+1} - H_i^j)} \tag{28}$$

At time t_i the temperature at node i is T_m. The temperature at the other node point follows from the linear interpolation

$$T_k^{j+\chi} = \chi(T_k^{j+1} - T_k^i) + T_k \,, \quad (k \ne i) \tag{29}$$

Combination of equations (27-29) in a control volume enthalpy solution result in an accurate and simple means of analysing a phase change. The accuracy of the technique may be observed on reference to Table 2 where results for the test

TABLE 2. SCHEME 1 RESULTS

Time hours	Position of Boundary Numerical	True	Temperature at .5m Numerical	True
13.08	.125	.125147	1.76680	1.78867
52.43	.25	.250544	.84987	.88075
117.89	.375	.375706	.30599	.32211
209.44	.5	.500767	0.0	.01427
327.06	.625	.625782	-1.96839	-1.91154
470.82	.75	.750821	-3.27352	-3.21357
641.09	.875	.876132	-4.20720	-4.16069
838.97	1.00	1.002260	-4.91175	-4.88205

problem using an explicit scheme with $\delta t = 900$ and $\delta x = .0625$ are compared with analytical values.

Scheme 2: Node jumping. A problem with the Scheme 1 is that the phase front will only coincide with a node point at specific points in time. Therefore results are only accurate at a very limited number of time steps. This will create difficulties in problems which involve temperature dependent properties. An obvious way of overcoming the drawbacks is to employ a variable time step such that the phase front is always on a node point thus ensuring that the temperature profiles are correct at all points in the numerical solution. In adopting this approach, however, an implicit scheme will need to be used. The implicit control volume method defined in equations (18) and (20) is well suited to such a "node jumping" application.

To illustrate the method of node jumping the freezing problem defined in Table 1 is used. If at time t the freesing front is on node p the nodal latent heats are

$$
\Delta H_i^j = \begin{cases} 0 & \text{if} & i < p \\ L/2 & \text{if} & i = p \\ L & \text{if} & i > p \end{cases}
$$

After a time step δt if the boundary has moved to the node point $p+1$ the nodal latent heats will be

$$
\Delta H_i^{j+1} = \begin{cases} 0 & \text{if} & i < p+1 \\ L/2 & \text{if} & i = p+1 \\ L & \text{if} & i > p+1 \end{cases}
$$

Therefore in this time step an amount $L/2$ of heat has been evolved from control volumes p and $p+1$ respectively. Hence for time step δt the source term in the implicit solution will be

$$
S_i^{j+1} = \begin{cases} 0 & i \neq p, \quad i \neq p+1 \\ L/(2C) & \text{otherwise} \end{cases}
$$

The only unknown in the implicit solution, therefore, is the value of time step, δt, required for the freezing front to move from $x = p\delta x$ to $x = (p+1)\delta x$. The application of the node jumping techniques is as follows. The values of S_p^{j+1} and S_p^i are set to $L/2$ with the remaining source components set to zero. Then a choice for the time step δt is made. With known $\underline{S}^{j+1}$ and δt the system of equations (18) become linear and may be solved by the Thomas algorithm [29]. Clearly the central problem is selecting the correct value for δt. When the correct value for δt is used in equation (18) the predicted nodal temperature $T_{p+1}^{j+1} = T_m$. This fact is used in order to develop a procedure for iterating towards the correct value of δt. An initial guess $[\delta t]_o$ would be the time step required for the boundary to jump from node p - 1 to node p. On calculating the temperature field from a solution of equation (18) a revised value follows from

$$[\delta t]_1 = [\delta t]_o + \frac{(T_m - [T_p^{j+1}]_o)[\delta t]_o}{[T_p^{j+1}]_o - L/2C - T_p^j} \tag{30}$$

On subsequent calculations the time step is modified as

$$[\delta t]_k = [\delta t]_{k-1} + \frac{(T_m - [T_p^{j+1}]_{k-1})([\delta t]_{k-1} - [\delta t]_{k-2})}{([T_p^{j+1}]_{k-1} - [T_p^{j+1}]_{k-2})} \tag{31}$$

This process continues until the value of δt is such that

$$\left| [T_p^{j+1}]_k - T_m \right| < \gamma$$

where γ is a convergence factor.

The node jumping method incorporated with the implicit enthalpy method, as outlined above, is very efficient. Test

TABLE 3. NODE JUMPING RESULTS

Time	Position of Boundary		Temperature at .5m	
Hours	Numerical	True	Numerical	True
12.97	.125	.124632	1.79821	1.79216
52.04	.25	.249622	0.87151	0.886201
117.18	.375	.374573	0.32077	0.325863
208.48	.5	.499617	0.0	0.00077
325.90	.625	.624673	-1.89248	-1.89781
469.42	.75	.749707	-3.19272	-3.20379
639.05	.875	.874734	-4.1371	-4.15167
834.78	1.00	.99976	-4.85315	-4.86946

problem 1 has been solved using this approach with $\delta x = 0.0625$. Results are compared with analytical values in Table 3. These results compare favourably with explicit solution results, Table 2, but required less than 3% of the CPU time on a DEC 11/34.

Note that in order to account for the boundary discontinuity in this problem for the first jump from $x = 0$ to $x = .0625$ the implicit method was applied with $\theta = 1$ (fully implicit) whilst in all subsequent jumps $\theta = \frac{1}{2}$ (Crank-Nicolson).

Scheme 3: Continuous track. The node jumping scheme only provides predictions for the position of the phase front when it is on a node point. On employing equation (24) a scheme may be developed which can accurately track the phase front over fixed time steps. If the ith control volume is changing state then the portion of the volume which has changed state will be given by equation (24) and the position of the phase boundary will be

$$X(t) = \delta x(i \pm S - \tfrac{1}{2}) \qquad (32)$$

where the sign depends on the nature of the problem in question. Application of this scheme simply involves the coding of equation (32) into the enthalpy finite difference scheme, equation (8). Such applications may produce accurate predictions for the movement of the boundary which compare well with more complex phase change front tracking techniques. An application of equation (32) is made in the solution of a freezing problem in the half space $x \geqslant 0$ with the thermal data of Table 4.

TABLE 4.   TEST PROBLEM 2

$T_m = 0$ $T_o = -20$ $T_I = 10$

$C_s = 1.76$ $C = 4.226$ $K_s = 2.22$ $K_\ell = .556$ $L = 338$

Predictions using an explicit scheme with $\delta x = 0.25$ and $\delta t = .002$ are compared with the analytical solution and predictions from three alternative schemes published by Furzeland [9] in Table 5. The solution from the enthalpy scheme gives the same order of accuracy as the other front tracking techniques but far less coding is required. Furthermore all the results presented by Furzeland were produced using starting solutions; the enthalpy solution uses none.

TABLE 5.   FRONT TRACKING RESULTS

Time	Furzeland			Eq.(32)	Exact
	(1)	(2)	(3)		
.0024	.0024	.0196	.0228	.0217	.0226
.0036	.0276	.0256	.0279	.0275	.0278
.0180	.0618	.0617	.0619	.0624	.0619
.0720	.1238	.1236	.1237	.1233	.1238
.1440	.1750	.1749	.1750	.1747	.1750
.2880	.2476	.2474	.2474	.2471	.2475

6.3 Non Cartesian Applications

All the remedial schemes outlined so far are, after
slight modification, applicable to problems in cylindrical
and spherical geometries. With symmetric boundary conditions
equation (24) which relates the portion of a control volume,
S, that has changed state to the nodal latent heat ΔH_i is
still valid. Due to the shape of the control volumes, a section of an annular ring in cylindrical co-ordinates, or an
element of a spherical shell in spherical co-ordinates the
value of S, when the phase front coincides with the node
point is

$$S = \tfrac{1}{2} \pm \begin{cases} \delta r/8r & , \text{ cylinder} \\ 3r\delta r/(12r^2+\delta r^2) & , \text{ sphere} \end{cases} \tag{33}$$

The corresponding value of the nodal latent heat is

$$\Delta H_i = \begin{cases} L(1-S) & , \quad \text{freezing} \\ LS & , \quad \text{melting} \end{cases} \tag{34}$$

This is the value that should be used in applications of the
"latent heat at node" and "node jumping" schemes. The equivalent form of the continuous tracking algorithm, equation (32)
in polar co-ordinates is

$$R(t) = \delta r(i + \alpha/2) \tag{35}$$

where α is evaluated from the following equations

$$(1 \pm \alpha)(\tfrac{1}{2} \pm (1 \pm \delta)\, r/8r) = S$$

for cylindrical coordinates and

$$\frac{3r^2 \delta r}{2}(1 \pm \alpha) \pm \frac{3r\delta r^2}{4}(1 - \alpha^2) + \frac{\delta r^3}{8}(1 \pm \alpha^3) = S(3r^2 \delta r + \frac{\delta r^3}{4})$$

for spherical co-ordinates

The "latent heat at node" scheme has been applied to the problem of inward solidification of a circular cylinder of unit radius. The thermal data is given in Table 6.

TABLE 6. CYLINDER PROBLEM

$$T_I = T_m = 1 \qquad T_o = 0$$

$$C = 1 \qquad K = 1 \qquad L = 1$$

In Table 7 results are compared with those obtained by Tao [47] who used an alternative technique

TABLE 7. CYLINDER RESULTS

	Enthalpy $\delta r = .025$ $\delta t = .00025$	Tao[47] $\delta a = .02$ $\delta t = .0002$
Position of Boundary	Time	Time
0.9	.00625	.0064
0.8	.02442	.02486
0.7	.05348	.05411
0.6	.09178	.09264
0.5	.13796	.13886
0.4	.18945	.19080
0.3	.24471	.24618
0.2	.29954	.30168
0.1	.34971	.35228
0.0	.38028	.38647

6.4 Mushy Region Solutions

In many physical problems the phase change does not occur at a single temperature but over a temperature half range ε. Therefore in the solution domain there is a region which exists in both the liquid and solid state simultaneously, the so called "mushy" region. Numerical enthalpy solutions of a mushy region problem are simply accommodated by assigning a value for ε in the temperature enthalpy relationship, equations (6) and (7).

In a direct application of the enthalpy method a mushy region is often advantageous because oscillations are greatly damped. This behaviour has been fully investigated by Bonacina et al [18] and Voller et al [43].

Of interest in the solution of problems involving a mushy region would be its extent. This poses a problem in that defined boundary conditions on the mushy region are not known. Therefore in a control volume which is changing state it becomes difficult to differentiate between the solid, mushy and liquid regions. One approach in tracking the movement of the phase change region would be to interpolate in temperature for the positions of $T_m + \varepsilon$ and $T_m - \varepsilon$. However, oscillations in evaluated temperature fields may still be expected. This in turn will lead to oscillations in predictions for the movements of the $T_m + \varepsilon$ and $T_m - \varepsilon$ isotherms. An alternative technique, in line with the remedial schemes presented above, would be to record the times at which nodal enthalpies take the values $C(T_m + \varepsilon) + L$ and $C(T_m - \varepsilon)$. These values indicate, in a freezing problem, that the phase change region is just entering or just leaving the control volume. A mushy region tracking using this approach ($\delta x = .0625$ $\delta t = 7200$) for test problem 1 when $\varepsilon = 1$ is illustrated in Figure 5.

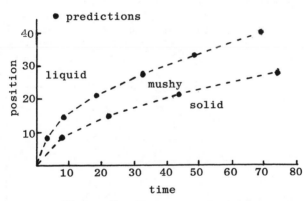

Figure 5. Movement of Mushy Region

7. TWO DIMENSIONAL APPLICATIONS

The problems of direct enthalpy methods will still be present in applications to two dimensional problems. Temperature histories will still exhibit step like oscillations. The position of the boundary between the solid and liquid regions will be located on the surfaces of the control volumes which are changing state. In addition the movement of this boundary will be discrete jumping from one location to the next as the control volume completes its phase change.

Many of the remedial enthalpy schemes developed for one dimensional problems cannot be generalised to more dimensions. Limited application of the interpretation of enthalpy, made in section 5, may be used however, to provide some elementary techniques to smoothly track the phase front movement. To examine possible applications the problem of freezing in a square cavity, where the length of the side is 2, is introduced. The thermal conditions for this problem are identical to those of test problem 1, see Table 1. Note that applications of a direct enthalpy method to a similar problem has been made by Crowley [48].

The standard method of tracking a phase front in two dimensions is to position it such that control volumes which are chaning state or are about to change state are separated from the control volumes which have changed state. In some applications, especially with a fine grid, such an approach may be sufficient. Morgan [49] for example, successfully uses this method in a conduction/convection freezing problem. The drawbacks are: 1) the boundary encloses an area which may be greater than the true liquid area, 2) the boundary profile is not smooth, and 3) the boundary movement is discrete. A more accurate division between the solid and liquid regions may be obtained on using equation (24) to "draw" a line separating solid and liquid within each control volume which is changing state. Then the sum of all the lines will make up the required boundary. This techniques has been used in the square cavity problem using the implicit enthalpy method with $\delta x = \delta y = 0.1$ and $\delta t = 5$ hours Predicted results for the position of the boundary in a quarter section

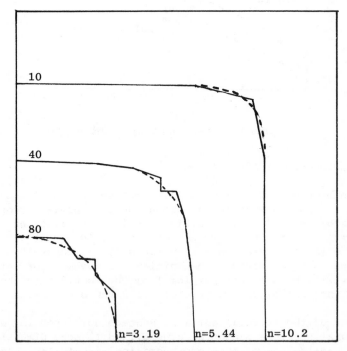

Figure 6. Position Of Boundary In Freezing Cavity

are shown in Figure 6. The boundary now splits the solid/liquid accurately but does not have a smooth profile. In addition it is difficult to automatically generate a boundary of this nature.

The solid lines in Figure 6 may be assumed to be close to the "true" position of the solid/liquid interface. Therefore what is needed is a means of smoothly interpolating the plots. To do this effectively some physical criteria is required. Equation (25) which relates the area which has changed state to the nodal latent heat values provides such a criteria. The object is to find an analytical expression for the phase front position which will correctly divide the total area into solid and liquid portions according to equation (25) and also closely interpolate the solid line plots in Figure 6. For the square cavity problem it is reasonable to postulate that the shape of the boundary is a square near to the surface reducing to a circle at the centre. The family of curves

$$x^n + y^n = r^n \qquad 2 \leq n \leq \infty \qquad (36)$$

exhibit this behaviour taking the shape of a circle of radius r when n = 2 and the shape of a square of side 2r as n becomes large. Therefore such curves would seem a suitable choice for the required interpolation. A possible procedure is as follows. At the time step in question the portion of the area still in the liquid state is estimated as $1 - S^*$, see equation (26). The value of r in equation (36) may be estimated on taking the control volume prediction of the freezing from along an axis. The required value of n for the interpolating plot then follows from solution of the equation

$$1 - S^* = \int_0^r (r^n - x^n)^{1/n} \, dx$$

or more simply the equation

$$\frac{1 - S^*}{r^2} = \int_0^1 (1 - \eta^n)^{1/n} \, d\eta \qquad (37)$$

A search routine may be employed to find the value of n with the RHS of equation (37) numerically integrated at each stage. Typical interpolating plots are shown as dashed lines in Figure 6. These plots are close to the control volume plots and more importantly the area cut by the plots agrees with the numerical predictions of equation (25).

The technique outlined above for tracking the boundary in two dimensions may appear to be limited. Voller and Shadabi [50], however, have developed a general methodology for applying similar techniques. They demonstrate this methodology by smoothly tracking a freezing front in an asymmetric region. In this application cubic spline interpolation is used in conjunction with equation (25) to produce analytical plots which correctly split the region of interest into solid and liquid fractions.

8. CONCLUSIONS

Enthalpy methods for solution of phase change problems have enjoyed wide application. Their main asset is that in a solution the conservation of heat may be satisfied without reference to the exact position of the phase change interface. This fact makes enthalpy methods very appealing in numerical solutions because problems may be solved on a fixed grid.

Basic drawbacks in numerical enthalpy methods have been documented. [18, 22, 43]. In this chapter a number of schemes, based on control volume techniques, have been proposed. These schemes by-pass these drawbacks and lead to accurate enthalpy solutions for one dimensional Stefan problems.

Applications of the enthalpy schemes presented in two-dimensional problems is not straight forward. By nature, however, two dimensional systems tend to be complex and an inadequacy to account for the position of the boundary may not represent a sensitive input to the system model.

The treatment of mushy regions represents another problem area. Even though the introduction of a mushy region tends to "blur" the numerical problems associated with direct enthalpy methods a detailed analysis of the region itself may prove difficult. This problem becomes important in modelling segregation in a solidification process where the size and extent of the mushy region is of prime interest [51].

The aim of this chapter has been to summarise control volume enthalpy techniques, demonstrate their general utility and provide pointers to areas where further work is required. In general enthalpy methods appear to offer useful tools in the analysis of phase change. The time is right for the development of numerical enthalpy techniques based upon both finite element and control volume approaches which may be incorporated into generalized phase change computer software.

REFERENCES

1. SZEKELY, J. and THEMELIS, N. J., Rate Phenomena in
   in Process Metallurgy, Wiley-Interscience, 1971.

2. CRANK, J. and GUPTA, R. S., A Moving Boundary Problem
   Arising From The Diffusion Of Oxygen In Absorbing
   Tissure. J. Inst. Maths Applics., Vol. 10, pp
   19-33, 1972.

3. TARZIA, D. A., Una Revision Problemas De Frontera Movil
   Y Libre Para La Eduacion Del Calor El Problema De
   Stefan. Mathematical Notae Universiclacl Nacional
   de Rosario, Argentina), Vol. 29, pp 147-241, 1981/82.

4. CARSLAW, H. S. and JAEGER, J. C., The Conduction Of Heat
   In Solids, Oxford University Press, 1959.

5. CRANK, J. The Mathematics Of Diffusion, Clarendon Press,
   Oxford, 1975.

6. FOX, L., What Are The Best Numerical Methods? Moving
   Boundary Methods In Heat Flow And Diffusion. Ed.
   Ockendon, J. R. and Hodgkins, W. R., Oxford Uni-
   versity Press, 1975.

7. CRANK, J., Finite Difference Methods, Moving Boundary
   Methods In Heat Flow And Diffusion. Ed. Ockendon,
   J. R. and Hodgkins, W. R., Oxford University Press,
   1975.

8. CRANK, J., How To Deal With Moving Boundaries In Thermal
   Problems, Numerical Methods In Heat Transfer, Ed.
   Lewis, R. W., Morgan, K. and Zienkiewicz, O.C.,
   Wiley-Interscience 1981.

9. FURZELAND, R. M., A comparative study of Numerical Methods
   In Moving Boundary Problems, J. Inst. Maths. Applics.
   Vol. 26, pp 411-429, 1980.

10. PATANKAR, S. V., Numerical Heat Transfer And Fluid Flow,
    Hemisphere, New York, 1980.

11. EYRES, N. R., HARTREE, D. R., INGHAM, J., JACKSON, R.,
    SARJANT, R. J., and WAGSTAFF, J. B., The Calculation
    Of Variable Heat Flow In Solids. Phil. Trans. R.
    Soc., Vol. 240 (A), pp 1-57, 1946.

12. SHAMSUNDAR, N. and SPARROW, E. M., Analysis of Multi-
    dimensional Conduction Phase Change Via The Enthalpy
    Model, J. Heat Transfer, Vol. 97(C), pp 333-340,
    1975.

13. ALBASINY, E. L., The Solution Of Non-Linear Heat Conduction Problems On The Pilot Ace. Proc. I.E.E., Vol. 103(B), pp 158-162, 1956.

14. LOCKWOOD, F. C., Simple Numerical Procedures For The Digital Computer Solution Of Non-Linear Transient Heat Conduction With Change Of Phase, J. Mech. Eng. Sci. Vol. 8, pp 259-263, 1966.

15. LONGWORTH, D., A Numerical Method To Determine The Temperature Distribution Around A Moving Weld Pool, Moving Boundary Methods In Heat Flow And Diffusion, Ed. Ockendon, J. R. and Hodgkins, W. R., Oxford University Press, 1975.

16. SZEKELY, J. and LEE, R. G., The Effect Of Slag Thickness On Heat Loss From Ladles Holding Molten Steel. Trans. Met. Soc., Trans AIME, Vol. 242, pp 961-965, 1968.

17. MEYER, G. H., Multidimensional Stefan Problems, SIAM J. Numer. Anal., Vol. 10(3) pp 353-366, 1974.

18. BONACINA, C., COMINI, G., FASANO, A. and PRIMICERIO, M., Numerical Solution of Phase Change Problems, Int. J. Heat Mass Transfer, Vol. 16, pp 1825-1832, 1973.

19. ROSE, M. E., On The Melting Of A Slab, SIAM J. Appl. Math., Vol. 15, pp 495-504.

20. ATTHEY, D. R., A Finite Difference Scheme For Melting Problems, J. Inst. Math. Appl., Vol. 13, pp 353-366, 1974.

21. ATTHEY, D. R., A Finite Difference Scheme For Melting Problems Based On The Method Of Weak Solutions, Moving Boundary Methods In Heat Flow And Diffusion. Ed. Ockendon, J. R. and Hodgkins, W. R., Oxford University Press, 1975.

22. DUSINBERRE, G. M., Numerical Methods For Transient Heat Flow, Trans ASME, Vol. 67, pp 703-710, 1945.

23. PRICE, P. H. and SLACK, M. R., The Effect Of Latent Heat On Numerical Solutions Of The Heat Flow Equation, Brit. J. Appl. Physics, Vol. 5, pp 285-287, 1954.

24. BAXTER, D. C., The Fusion Times Of Slabs And Cylinders, J. Heat Transfer, Vol. 84(C), pp 317-326, 1962.

25. KAMENOMOSTSKAJA, S. L., On Stefans Problem. Mat. Sab. Vol. 53, pp 489-514, 1961.

26. OLEINIK, O. A., A Method Of Solution Of The General
    Stefan Problem, Sov. Math. Dokl., Vol. 1, pp 1350-
    1354, 1960.

27. ELLIOTT, C. M. and OCKENDON, J. R., Weak And Variational
    Methods For Moving Boundary Problems, Pitman,
    London, 1982.

28. SOLOMON, A., Some Remarks On The Stefan Problem, Math.
    Comp., Vol. 20, pp 347-360, 1976.

29. SMITH, G. D., Numerical Solutions Of Partial Differential
    Equations: Finite Difference Methods. Oxford Uni-
    versity Press, 1978.

30. WHITE, R. E., An Enthalpy Formulation Of The Stefan
    Problem, SIAM J. Numer. Anal., Vol. 19(6), pp 1129-
    1157, 1982.

31. WHITE, R. E., A Numerical Solution Of The Enthalpy
    Formulation Of The Stefan Problem, SIAM J. Numer.
    Anal., Vol. 19(6), pp 1158-1172, 1982.

32. VOLLER, V. R., Implicit Finite Difference Solutions Of
    The Enthalpy Formulation Of Stefan Problems. Sub-
    mitted To IMA J. Numer. Anal. (1984).

33. WOOD, A. S., RITCHIE, S. I. M., and BELL, G. E., An
    Efficient Implimentation Of The Enthalpy Method,
    Int. J. Num. Meth. Eng. Vol. 17, pp 301-305, 1981.

34. GOURLAY, A. R., Hopscotch A Fast Second Order Partial
    Differential Equation Solver. J. Inst. Maths.
    Applics., Vol. 6, pp 375-390, 1970.

35. COMINI, G., DEL GUIDICE, S., LEWIS, R. W. and
    ZIENKIEWICZ, O. C., Finite Element Solutions Of
    Non-Linear Heat Conduction Problems With Special
    Reference To Phase Change. Int. J. Num. Meth. Eng.
    Vol. 8, pp 613-624, 1974.

36. MORGAN, K. LEWIS, R. W., and ZIENKIEWICZ, O. C., An
    Improved Algorithm For Heat Conduction Problems With
    Phase Change. Int. J. Num. Meth. Eng. Vol. 12,
    pp 1191-1195, 1978.

37. LI, C. H., A Finite-Element Front Tracking Method For
    Stefan Problems, IMA J. Numer. Anal. Vol. 3,
    pp 87-107, 1983.

38. ROLPH, W. D. and BATHE, K. J., An Efficient Algorithm
    For Analysis Of Non-Linear Heat Transfer With Phase
    Changes. Int. J. Num. Eng., Vol.18, pp 119-134,1982.

274

39. WINZELL, B., Finite Element Galerkin Methods For Multi-
    Phase Stefan Problems. Appl. Math. Modelling,
    Vol. 7, pp 329-344, 1983.

40. RONEL, J. and BALIGA, B. R., A Finite Element Method
    For Unsteady Heat Conduction In Materials With Or
    Without Phase Change. ASME 79-WA/HT-54, 1979.

41. VOLLER, V. R., Interpretation Of The Enthalpy In A
    Discretized Multi-Dimensional Region Undergoing A
    Melting/Freezing Phase Change, Int. Comm. Heat Mass
    Transfer, Vol. 10(4), pp 323-328, 1983.

42. VOLLER, V. R. and CROSS, M., An Explicit Method To Track
    A Moving Phase Change Front. Int. J. Heat Mass
    Transfer, Vol. 26(1), pp 147-150, 1983.

43. VOLLER, V. R., CROSS, M. and WALTON, P. G., Assessment
    Of Weak Solution Numerical Techniques For Solving
    Stefan Problems, Numerical Methods In Thermal
    Problems, Ed. Lewis, R. W. and Morgan, K., Pineridge
    Press, Swansea, 1979.

44. BELL, G. E., On The Performance Of The Enthalpy Method,
    Int. J. Heat Mass Transfer, Vol. 25, pp 587-589,
    1982.

45. VOLLER, V. R. and CROSS, M., Accurate Solutions Of Moving
    Boundary Problems Using The Enthalpy Method, Int. J.
    Heat Mass Transfer, Vol. 24, pp 545-556, 1981.

46. VOLLER, V. R. and CROSS, M., Estimating The Solidifi-
    cation/Melting Times Of Cylindrically Symmetric
    Regions. Int. J. Heat Mass Transfer, Vol. 24,
    pp 1457-1462, 1981.

47. TAO, L. N., Generalized Numerical Solutions Of Freezing
    Of Saturated Liquid In Cylinders And Spheres.
    A.I.Ch.E. Journal, Vol. 13, pp 165-169, 1967.

48. CROWLEY, A. B., Numerical Solution Of Stefan Problems.
    Int. J. Heat Mass Transfer, Vol. 21, pp 215-219,
    1978.

49. MORGAN, K., A Numerical Analysis Of Freezing And Melting
    With Convection. Com. Meth. App. Mech. Eng.,
    Vol. 28, pp 275-284, 1981.

50. VOLLER, V. R. and SHADABI, L., Enthalpy Methods For
    Tracking A Phase Change Boundary In Two Dimensions.
    To appear in Int. Comm. Heat Mass Transfer 1984.

51. MOORE, J. J., SHAH, N. A. and VOLLER, V., The Control of
    Channel Segregation In Cast Steel - A Review.
    American Foundarymen's Society, Paper No.83-08, 1983.

CHAPTER 11

NONLINEAR COUPLED TRANSIENT HEAT CONDUCTION
AND QUASI-STEADY ELECTRIC CURRENT FLOW

Struan R. Robertson

Mechanical Engineering Department
University of Lowell
Lowell, MA 018 54

ABSTRACT

Heat conduction problems that involve Joule heating,
where the temperature excursions are large, are
highly nonlinear. This is due to the temperature
dependence of both the thermal and electrical
properties of materials. The electric field problem
is coupled to the heat conduction problem in this
case. For quasi-steady currents, magnetic field
effects can be ignored. In this case the equations
for the two fields are mathematically analogous so
that the same formulation can be used for both sets
of equations. The finite element method is applied
to the problem with the effects of phase change
being included in the formulation. The finite
difference method can be applied in a similar
fashion.

INTRODUCTION

Many particularly difficult heat transfer
problems arise in electric power handling equipment
and devices as well as certain solid state devices.
The difficulty is twofold. The heat conduction
problem can be nonlinear because of the temperature
dependent behavior of the thermal properties.
Further, if the temperature excursions are large
enough, the electrical resistivity will be
temperature dependent. Thus, the material is heated
internally due to Joule heating while the electrical
behavior depends on the resulting temperature
field. This causes coupling of the two fields.

The finite element method provides an excellent
vehicle for the analysis of such problems. The
quasi-steady electric current problem has an
equation that is mathematically analogous to that

for steady state heat conduction. Therefore, the same formulation that is used for transient heat conduction can be readily adapted for use in quasi-steady current problems. Further, the coupling that occurs in nonlinear problems can be dealt with by suitably modifying an existing nonlinear heat conduction code.

Traditionally, problems that involved Joule heating have been treated in an uncoupled manner. Two examples of this using finite elements appear in references [1] and [2]. Coupled problems have been treated only for the simplest cases though one interesting article dealing with thermal switching appears in reference [3]. The program that was developed to solve this class of problems is applied to several simple problems for which other solutions exist and two complex problems with no known solutions. The last problem, in particular, demonstrates a situation where only a fully coupled nonlinear analysis can give meaningful results.

GOVERNING EQUATIONS

The equation of heat conduction for an isotropic material with spatially varying properties [4], using index notation, is

$$(kT_{,i})_{,i} + Q = \rho c \dot{T} \qquad \text{in } D \qquad (1)$$

with the boundary and initial conditions

$$kT_{,n} = f1(T,x,t) \qquad \text{on } S1 \qquad (2)$$

$$T = f2(x,t) \qquad \text{on } S2 \qquad (3)$$

$$T(x,0) = f3(x) \qquad \text{in } D \qquad (4)$$

The electric field equation for steady or slowly varying current flow with spatially varying conductivity is [5]

$$(\sigma V_{,i})_{,i} - q = 0 \qquad \text{in } D \qquad (5)$$

with the boundary conditions

$$\sigma V_{,n} = g1(x,t) \qquad \text{on } SI \qquad (6)$$

$$V = g2(x,t) \qquad \text{on } SII \qquad (7)$$

The subscript (,n) denotes differentiation in the direction normal to the surface. D is the interior domain of the body and S the boundary of D. S1 + S2

= S and SI + SII = S but S1 need not coincide with SI nor S2 with SII. V is the electric potential in volts, q the time rate of change of charge density in A/s-cc, σ the electrical conductivity in 1/ohm-cm, g1 the applied surface current in A/cm^2 and g2 the applied surface potential in volts.

Both thermal and electrical properties, k,c and σ, are assumed to be temperature dependent.

The Joule heating [5,6,7] is determined by the following equations.

$$E_i = -V_{,i} \qquad\qquad (8)$$

$$J_i = \sigma E_i \qquad\qquad (9)$$

$$P = J_i E_i = \sigma V_{,i} V_{,i} \qquad\qquad (10)$$

where E_i is the electric field in volts/cm and is analogous to the temperature gradient, J_i is the current density in amps/cm^2 and is analogous to heat flux while P is the Joule heating in watts/cc. P, therefore, contributes to the internal heat generation function Q in eq. (1).

FINITE ELEMENT EQUATIONS

Comparing eqs. (1-3) with eqs. (5-7) shows, in the absence of transient thermal effects, that they are mathematically equivalent. Thus, the formulation of the finite element equation for the electric field is the same as for the temperature field. The formulation for the thermal equation is accomplished using Galerkin's method [8] which is well known and will not be repeated here. At this point it is worth noting that a finite difference approach would proceed along similar lines. The finite element equation for heat conduction is

$$[K]\underline{T} + [C]\underline{\dot{T}} - \underline{Q} = 0 \qquad\qquad (11)$$

and for the electric field it is

$$[A]\underline{V} + \underline{B} = 0 \qquad\qquad (12)$$

where [K] is the thermal conductivity matrix, [C] the thermal heat capacity matrix, $\underline{T}$ the nodal temperature vector, $\underline{Q}$ the thermal load vector, [A] the electrical conductivity matrix, $\underline{V}$ the nodal voltage vector and $\underline{B}$ the current load vector. [K], [C] and [A] depend on $\underline{T}$ in general and $\underline{Q}$ depends on $\underline{V}$ and $\underline{T}$ in general.

The heat load vector $\underline{Q}$ will depend on the Joule heating as given by eq. (10) as well as any other heat sources. Using a planar 8-node isoparametric element [8,9] for example, the voltage is given by

$$V(x,y) = \sum_{i=1}^{8} N_i(x,y)V_i \qquad (13)$$

where the V_i are nodal voltages and the $N_i(x,y)$ are the interpolation functions. The terms in the gradient of V are found from eq. (13) to be

$$V_{,x} = \sum N_{i,x}V_i$$
$$V_{,y} = \sum N_{i,y}V_i \qquad (14)$$

so that eq. (10) becomes

$$P = \sigma (V_{,x}^2 + V_{,y}^2)$$

or

$$P = \sigma [(\sum N_{i,x}V_i)^2 + (\sum N_{i,y}V_i)^2] \qquad (15)$$

The contribution of the Joule heating to the heat load vector for the element will be

$$\underline{Q} = \int_{-1}^{1} \int_{-1}^{1} wP\,N\,|J|\,dr\,ds \qquad (16)$$

where eq. (15) is used in eq. (16). $|J|$ is the Jacobian determinant of the transformation from x,y coordinates to r,s coordinates and w is the element's thickness. The integration is performed by Gaussian quadrature, the quantities within the integral being evaluated at the Gauss integration points [8,9].

TREATMENT OF MATERIAL PROPERTIES

Not only are the temperature dependence of the thermal and electrical properties of interest but also the latent heat effect is of concern. When latent heat effects (phase change) are not important the nonlinear behavior of the material properties is easily handled by a straight forward look up with linear interpolation in a given table of values. Phase change quite often takes place at a fixed temperature and so presents a problem with a sharp moving boundary. In other cases, the change of phase may occur over a temperature range, in which case the moving boundary is smeared over some

distance. One manner of treating the latent heat
effect in numerical modeling is to include it in the
specific heat by defining a fictive zone of phase
change wherein the specific heat is given a large
value such that the additional area under the curve
is equal to the latent heat. This is illustrated in
Fig. 1. In the case of materials that change phase
at one temperature the zone is made very narrow.

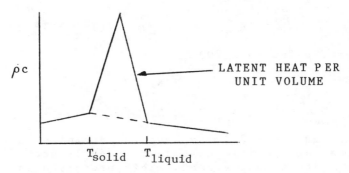

FIGURE 1. Variation of specific heat per unit
volume with temperature including phase change.

In the case of materials that change phase over a
range of temperature the zone is from the solidus to
the liquidus temperature. It is cumbersome to deal
directly with the curve shown in Fig. 1 because of
the sharp jump. The enthalpy method, however,
provides a straight forward means of solving phase
change problems when used with the finite element
method [10,11]. The enthalpy, relative to some base
temperature, is found by integrating ρc with respect
to temperature thus

$$H(T) = \int_{T_0}^{T} \rho c(s)\,ds \qquad (17)$$

The value of ρc at any temperature is then just the
slope of $H(T)$,

$$\rho c = dH/dT \qquad (18)$$

As it stands, this does not appear very useful.
However, in the evaluation of the specific heat
matrix, an average value of the specific heat that
properly accounts for the latent heat effect is
required at each Gauss integration point. This is
accomplished by taking the derivative of $H(T)$ over
the element as follows. The temperature in the
element is evaluated in terms of the nodal values by

$$T = \sum_i N_i(x,y)T_i \qquad (19)$$

The enthalpy is evaluated in the same way

$$H = \sum N_i(x,y)H_i \qquad (20)$$

For a two dimensional problem, the derivative of H with respect to T in the direction of the temperature gradient is

$$(\rho c)_{avg} = dH/dT$$

$$= (H_{,x}T_{,x} + H_{,y}T_{,y})/(T^2_{,x} + T^2_{,y}) \qquad (21)$$

where, using eq. (20)

$$H_{,x} = \sum N_{i,x}H_i$$
$$H_{,y} = \sum N_{i,y}H_i \qquad (22)$$

Thus, eqs. (19-22) are used to determine an average value of specific heat that properly accounts for the latent heat effect. This procedure is also a good way of dealing with other highly nonlinear material properties.

The code takes temperature dependent data in tabular form. A tabular representation of the enthalpy function, for example, is automatically generated by integration of the specific heat data using linear interpolation between entries. This gives the i th enthalpy term in the table in terms of the i-1 term as

$$H_i = H_{i-1} + \rho c_{i-1}(T_i - T_{i-1}) +$$
$$(m_i/2)(T_i - T_{i-1})^2 \qquad (23)$$

where

$$m_i = (\rho c_i - \rho c_{i-1})/(T_i - T_{i-1}) \qquad (24)$$

It should be noticed that H varies parabolically between tabular entries. Thus, parabolic interpolation is used when determining the nodal values of H for use in eqs. (19-22), i.e.,

$$H = H_{i-1} + \rho c_{i-1}(T - T_{i-1}) + (m_i/2)(T - T_{i-1})^2 \qquad (25)$$

where

$$T_{i-1} < T < T_i$$

T is the temperature of the node and H is the corresponding enthalpy.

SOLUTION OF THE FINITE ELEMENT EQUATIONS

For transient problems the time derivative is approximated using Euler's backward difference. The system matrices are reformed at each time step based on the previously computed values of temperature. The electric field equation is solved first, followed by the computation of the Joule heating contribution to the heat load vector. The thermal problem is then solved and the process repeated until the solution is completed. In summary, at time t+dt

STEP 1 $[A]_t \underline{V}_{t+dt} = -\underline{B}_t$

STEP 2 Q using eq. (16)

STEP 3 $\left[[K]_t + (1/dt)[C]_t\right] \underline{T}_{t+dt} =$

$$(1/dt)[C]_t \underline{T}_t + \underline{Q}_t$$

For static problems the method of successive approximations is used together with loads that increase from zero to the final desired value. Iteration can be performed within each load step. In summary,

STEP 1 $[A]_i \underline{V}_{i+1} = -\underline{B}_i$

STEP 2 Q using eq. (16)

STEP 3 $[K]_i \underline{T}_{i+1} = \underline{Q}_i$

where the subscript i indicates the i th load step. Convergence is checked using the Euclidean norm [9], namely,

$$\|\Delta \underline{T}\| / \|\underline{T}\| = \left[\sum_{j=1}^{N} (T_{j,i} - T_{j,i-1})^2 \bigg/ \sum_{j=1}^{N} (T_{j,i})^2\right]^{1/2}$$

where N is the number of nodes in the model. If this norm is less than a given tolerance the solution is considered to have converged.

The same element subroutines are used for the electric conductivity matrix [A] as for the thermal conductivity matrix [K]. Further, the same assembly and solution schemes are used. The code that was written to implement this scheme is an in core program with 2000 degrees of freedom. The system matrix is stored as a column vector using a skyline procedure and the equations are solved by Gaussian elimination [8,9].

The method is applied to several examples. The first problem can be verified by an exact solution. The next two problems are compared with a finite difference model, for which a brief description is given. The last two problems examine situations for which no known solution is available.

EXAMPLE 1

An insulated cube 1 cm on a side and initially at 20 C is heated by a constant current. The heat generated per unit volume is

$$Q = I^2 r/4.184 \quad cal/cc$$

where I is the current and r is the electrical resistivity which is taken to be a linear function of temperature,

$$r = a + bT$$

The heat conduction problem reduces to the following simple equation

$$dT/dt = Q/\rho c = g(a + bt)$$

where $g = I^2/4.184\rho c$. The solution of the equation is

$$T = (a/b + T_o)\exp(gbt) - a/b$$

The finite element model is shown in Fig. 2. This is more complicated than is called for in this simple problem but the same model is used again for the next problem. Further, the planar element is used more to demonstrate its correctness than out of necessicity. The reason for using a four node element rather than an eight node one is that for many nonlinear problems the four node element behaves better. In fact, this author has found that when there is phase change it is necessary to introduce a ninth center node in an eight node element in order to eliminate spatial oscillations in the temperature field.

The right end is at zero volts and a current I is applied to the left end using a 2-node boundary element. The sides are electrically insulated. The thermal model is fully insulated.

The code uses electrical resistivity for

computations instead of electrical conductivity.
Since r is assumed to be linear only two data points
are needed for its representation. On the other
hand, if conductivity were used many more points
would be needed since $\sigma = 1/r$ which is a nonlinear
function of temperature. Taking a = 1.465xE-6 and
b = 6.63xE-9 gives r = 1.465 micro-ohm-cm at 0 C and
14.72 at 1000 C. The current is I = 300,000 amps.
Thermal properties are assumed constant. Thus the
thermal problem is linear but the electrical problem
is nonlinear. For copper at 20 C, k = 0.94
cal/cm-s-C and ρc = 0.86 cal/cc-C. The exact
solution is

$$T = 241\exp(165.8t) - 221$$

The voltage drop across the cube is

$$V = (a + bT)I = (0.4395 + 0.1989 \times 10^{-4}T)$$

Table 1 compares the exact and the finite element
solutions for the temperature, heat generation and
voltage drop time histories. The time step used in
the finite element analysis was t = 0.00005 s.

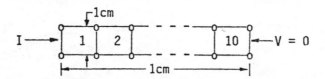

Figure 2. Model for Joule heating using 10 4-node
isoparametric elements 1cm thick.

TABLE 1						
	Exact solution			Finite element solution		
t s	T °C	Q 10^4 cal/sec	V volts	T °C	Q 10^4 cal/sec	V volts
0.00	20	3.4365	0.4793	20	3.436	-
.002	114.8	4.7886	0.6678	114.3	4.781	0.6613
.004	246.8	6.6711	0.9304	245.5	6.651	0.9200
.006	430.7	9.2947	1.296	428.0	9.253	1.280
.008	687.0	12.9502	1.806	681.9	12.870	1.781
.010	1044.0	18.0432	2.516	1035.0	-	2.477

EXAMPLE 2

In this problem both the thermal and electrical properties are temperature dependent. The problem is illustrated in Fig. 3. The left end is cooled by convection while the other sides are thermally insulated. The bottom is at 1 volt and the top at 0 volts with the ends electrically insulated. The current flow is normal to the heat flow. The mesh for the previous problem is used again. The material properties are given in Table 2.

A finite difference model was used to check the solution. This model consisted of the usual one dimensional finite difference representation of the heat flow equation together with a lumped electrical resistance at each node. The resistances were connected in parallel. Their values were updated at each time step to reflect their temperature dependence. Fig. 4 shows the finite difference network used.

The finite difference equation for boundary node 1 is

$$\rho_1 c_1 w(T_1 - TA_1)/dt - K_{1,2}(T_2 - T_1) - h(T_b - T_1) = wQ_1$$

while for an interior node it is

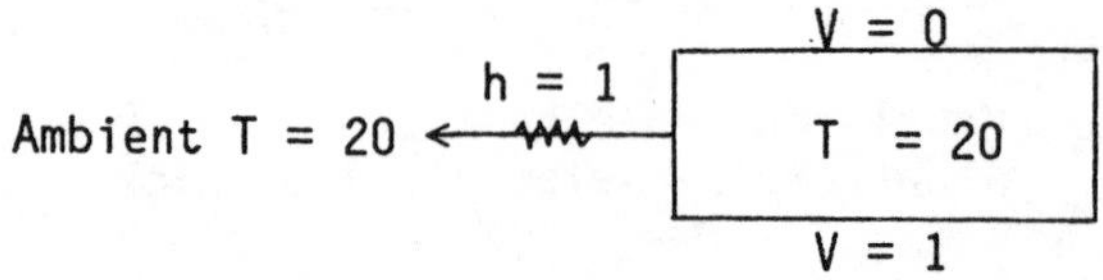

Ambient T = 20 ← $-\text{WW}-$ h = 1

V = 0

T = 20

V = 1

Figure 3. A convectively cooled, electrically heated cube

TABLE 2

T °C	k cal/cm-s-C	ρc cal/cc-C	r micro-ohm-cm
0	0.9	0.8	2.0
1000	0.6	0.8	8.0

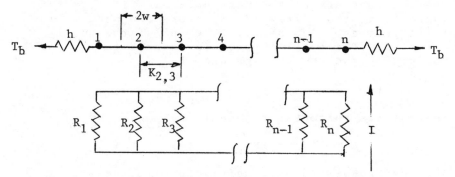

Figure 4. Showing both the thermal and electrical finite
difference networks.

$$2\rho_i c_i w(T_i - TA_i)/dt - K_{i-1,i}(T_i - T_{i-1}) -$$
$$K_{i,i+1}(T_{i+1} - T_i) = 2wQ_i$$

and for the boundary node n it is

$$\rho_n c_n w(T_n - TA_n)/dt + K_{n-1,n}(T_n - T_{n-1}) -$$
$$h(T_b - T_n) = wQ_n$$

where dt is the time step, w the half width of an
interior cell and the full width of a boundary cell,
TA_i is the temperature for the previous time step, T_b
is the bulk or ambient temperature and $K_{i,j}$ is the
conductance from node i to node j. Note that the
formulation is implicit.

The electrical resistance associated with each
node is

$$R_i = Lr_i/2wd$$

where r = a + bT, as in the first example, L is the
length of the model in the direction of current flow
(normal to heat flow) and d is the thickness of the
model. For a boundary node the 2 in the denominator
is dropped. Since the resistors are in parallel the
total resistance to current flow is

$$R = 1/\sum_i (1/R_i)$$

The current associated with each node is $I_i = v/R_i$ where v is the voltage drop across the model. The total current is simply the sum of the nodal currents. The total current is related to the total resistance and the voltage drop by

$$v = IR$$

so that the current for node i is

$$I_i = IR/R_i$$

The heat generation per unit volume for node i is

$$Q_i = (IR)^2/4.184R_i 2wLd \quad cal/s-cm^3$$

where the 2 is dropped from the denominator for boundary nodes.

The results for this problem are given in Table 3. The agreement between the finite element and the

	Finite difference		Finite element	
0.001s	cooled end	insulated end	cooled end	insulated end
2	198.2	236.1	195.2	235.8
4	286.3	390.1	279.4	389.5
6	342.0	516.6	332.1	515.7
8	381.9	626.5	370.1	625.3
10	412.8	725.4	399.5	723.6

TABLE 3
End temperatures in °C

finite difference model is quite good. Again, a time step of 0.00005 s was used.

EXAMPLE 3

This problem, shown in Fig. 5, is concerned with the electrical heating of a 3 layered laminate. The bottom is at 1 volt and the top is at 0 volts while the ends are electrically insulated. All surfaces are thermally insulated and the initial temperature is 20 C. The electrical resistivity and thermal conductivity vary linearly with temperature while the specific heat is constant. These are given in Table 4.

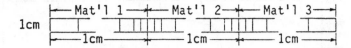

Figure 5. Model of an electrically heated laminate

TABLE 4									
	mat'l 1			mat'l 2			mat'l 3		
T	k	ρc	r	k	ρc	r	k	ρc	r
0	0.02	0.9	80.	0.9	0.8	2.0	0.04	0.9	80.
1000	0.06	0.9	100.	0.6	0.8	8.0	0.07	0.9	100.
The units are the same as those in Table 2.									

The center layer is a good thermal and electrical conductor
while the two outer layers are poor conductors. Steep
thermal gradients are to be expected at the interfaces
between layers. Thus, a variable mesh spacing is used. Ten
4-node elements are used in each layer for a total of 30
elements. The widths of the elements from the left are 0.2,
0.2, 0.2, 0.1, 0.1, 0.05, 0.05, 0.05, 0.025, 0.025 which
is at the end of the first layer then 0.025, 0.025, 0.05,
0.1, 0.2 cm which is at the middle of the 2nd layer. The
remaining elements' widths are in reverse order. The time
step was 0.00005 sec. The finite difference model used 10
equally spaced cells in each layer. The temperature
variations at the left end and center are given in Table 5.

TABLE 5				
	Fin. Diff.		Fin. Elem.	
t (0.001s)	T_{left}	T_{ctr}	T_{left}	T_{ctr}
2	26.6C	236.9C	26.5C	235.6C
4	33.2	391.2	33.0	389.0
6	39.8	517.9	39.5	514.9
8	46.3	627.9	46.0	624.2
10	52.9	726.5	52.5	722.0

The temperature distribution at time 0.01 sec is shown in
Fig. 6. The discrepancy at the interfaces is due
in part to the fact that the finite difference mesh had
equal spacing and was not representing the steep gradients
there.

290

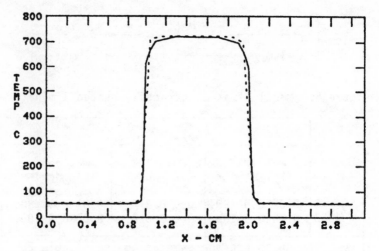

Figure 6.   Temperature distribution at 0.01 sec
            - - - - finite difference solution
            ———— finite element solution

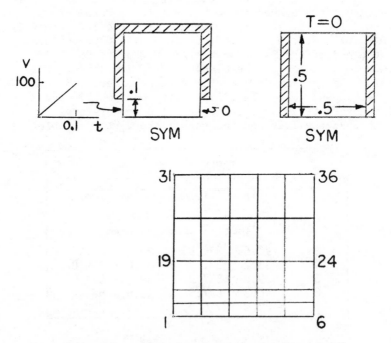

Figure 7.   Model of a thermal switch
            showing the boundary conditions.

EXAMPLE 4

This problem is concerned with thermal switching [3]. Thermal switching occurs in certain solid state devices which have very high electrical resistance below a certain threshold temperature and act like an open switch. Above the threshold temperature the electrical resistivity drops and the device acts like a closed switch. The device is shown in Fig. 7 along with the finite element mesh. As the applied voltage builds up, the Joule heating increases and the temperature will rise and the material will undergo a phase change like transformation after which the electrical resistivity drops significantly causing the device to act like a closed electrical switch. The thermal and electrical proerties are given in Table 6.

TABLE 6 properties for the thermal switch			
T	r	k	ρc
-10.0	0.001	1.0	1.0
10.0	0.001		1.0
11.0			500.0
12.0	0.00001		1.0
1000.0	0.00001	1.0	1.0

As shown in Fig. 7, the top is held at T = 0 while the bottom is a symmetry boundary. The sides are thermally insulated. The body is electrically insulated except for a portion of the vertical sides where current is permitted to flow. The voltage applied to the left electrode is a ramp that increases from 0 to 100 in 0.1 s. Fig. 8 gives the temperature time history for nodes 1 and 3. Note that as soon as the phase change occurs, the temperature begins to run away. This is due to the sudden decrease in electrical resistivity which permits more current to flow. Because Joule heating is proportional to voltage squared divided by resistance, the rate of heating increases in inverse proportion to the decrease in resistance for a fixed voltage. Fig. 9 shows the temperature distribution across the device just after switching has occurred.

292

Note that the x-axis, at T = 0, corresponds to the
top boundary of the model and that the bottom curve
corresponds to nodes 19 to 24. The slight waviness
of the curve is due to the facts that these nodes
are in the phase transition zone and that the mesh
is fairly crude. Once they have passed through the
phase transition zone the temperatures behave as
nodes 13 to 18 do.

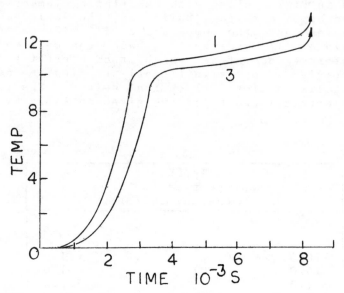

Figure 8. Temperature history of nodes
1 and 3.

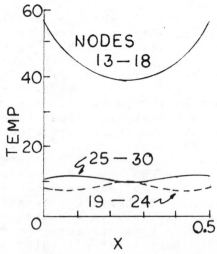

Figure 9. Temperature distribution just
after switching has occured.

EXAMPLE 5

This final problem concerns a copper disc into which current flows through a small central spot of radius 0.0118 cm. The back surface of the disc is at zero potential. The remainder is electrically insulated. The two dimensional axisymmetric model in Fig. 10 uses 336 nodes and 300 4-node isoparametric elements. The current flowing through the spot is a uniform pulse of 0.010 sec duration with an amplitude of 1450 amps. For the heat conduction model the boundaries are assumed to be

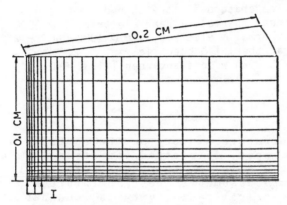

FIGURE 10. Axisymmetric finite element model showing the current source.

insulated because of the short duration of the event. A table of data for copper follows. Note the sharp increase in the resistivity and the drop in conductivity with transition through the phase change zone.

TABLE 7 properties of copper			
T (°C)	r (micro-ohm-cm)	k (cal/s-°C-cm)	ρc (cal/cc-°C)
0	1.5	0.94	0.82
200	1.7	0.93	-
540	-	0.814	-
1082.95	8.8	0.55	1.025	LATENT
1083.00	-	-	953.5	HEAT
1083.05	21.1	0.32	0.972	EFFECT
1200	22.1	-	-
2000	22.1	0.32	0.972

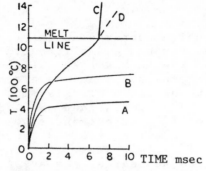

FIGURE 11. Temperature history at radius=0, z=0;
A - Nonlinear thermal, linear electric, r @ 200C
B-Linear thermal & electric, k,c @ 1082.95C, r @ 200C
C-Nonlinear thermal & electric
D-Same as C with r constant when T>1082.95C

The temperature history on the surface, at the center of the spot, is plotted for several cases in Fig. 11. Curve A is for the case where the thermal properties are assumed to be temperature dependent while the electrical resistivity is constant at its 200 C value. Thus case A is nonlinear thermally and linear electrically so they can be decoupled. Curve B is for the case where the thermal properties are constant at their 1082.95 C values (just prior to melting) and the resistivity is the same as in case A. Thus case B is linear and the thermal and electrical parts decouple. Curve C is for the fully nonlinear case where both the thermal and electrical properties are temperature dependent. In this case the problem is fully coupled. Note the large discrepencies between curves A,B and C.

An interesting thing to notice in curve C is the apparent instability once the melting point is reached on the surface. This is due to the jump in the value of resistivity as can be demonstrated in the following way. Consider the same case as C but not permitting the resistivity to exceed its 1082.95 C value even though temperature is increasing. The result is shown in curve D. Fig. 12 shows the surface temperature distribution prior to and after the onset of melting.

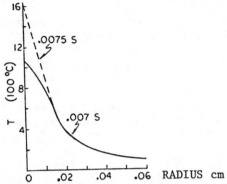

FIGURE 12. Surface temperature before and after melting

DISCUSSION

The first three problems dealt with cases for which there were other solutions. These give some confidence in the finite element code. The last two problems deal with situations where only a fully coupled nonlinear analysis will do. The last case, in particular, shows that using uncoupled analyses (A and B) one could predict many solutions simply by changing the temperatures at which the properties are to be evaluated and that the fully coupled nonlinear analysis C is the only one that predicts the behavior seen in practice.

ACKNOWLEDGEMENT

I wish to acknowledge the support given me by the people at GTE Research Laboratories, Waltham, Mass. where this work was carried out.

REFERENCES

1. Armor, A. F. - Transient, Three-Dimensional, Finite Element Analysis of Heat Flow in Turbine-Generator Rotors. IEEE Power Eng. Soc. Summer Meeting, Vancouver, B.C., 1979, Pap F 79, 719-6, Publ. IEEE, NY, 1979.

2. Robertson, S. R. - Finite Element Analysis of the Thermal Behavior of Contacts. IEEE Trans, Vol CHMT-5, No 1, pp3-10, Mar 1982.

3. Gelder, D. & Guy, A. G. - Current Problems in the Glass Industry. Moving Boundary Problems in Heat Flow and Diffusion, Eds. Ockendon, J. R. & Hodgkins, W. R., Clarendon Press, Oxford 1975.

4. Carslaw, H. S. & Jaeger, J. C. - Conduction of Heat in Solids, Oxford University Press, London 1959.

5. Abraham, M. & Becker, R. - The Classical Theory of Electricity and Magnetism, 2nd Ed., Hafner Publishing Co., NY 1950.

6. Landau, L. D. & Lifshitz, E. M. - Electrodynamics of Continuous Media, Pergamon Press, Oxford, 1960.

7. Pugh, E. M. & Pugh, E. W. - Principles of Electricity and Magnetism, 2nd Ed., Addison-Wesley, Reading, Mass. 1970.

8. Zienkiewicz, O. C. - The Finite Element Method, 3rd Ed., McGraw-Hill, London 1977.

9. Bathe, K. J. & Wilson, E. L. - Numerical Methods in Finite Element Analysis, Prentice-Hall, NJ 1976.

10. Comini, G. et al - Finite Element Solution of Nonlinear Heat Conduction Problems with Special Reference to Phase Change. Int. J. Num. Meth. Eng., Vol. 8, 613-624, 1974.

11. Comini, G. & DelGuidice, S. - Thermal Aspects of Cryosurgery. ASME J. Heat Transfer, Vol 98, No. 4, pp153-159, Mar, 1967.

CHAPTER 12

TRANSIENT PACKED BED HEAT TRANSFER: NUMERICAL SOLUTIONS.

PETER JOHN HEGGS

SENIOR LECTURER IN CHEMICAL ENGINEERING
LEEDS UNIVERSITY
LEEDS LS2 9JT., UK.

1. INTRODUCTION

The transient transfer of heat between a fluid flowing
through a packed bed occurs in many industrial and domestic
applications.   The overall transfer of heat is often a
combination of several competing and sequential mechanisms,
namely: fluid-to-particle convection, internal particle conduction, fluid conduction and effective particle conduction
in directions parallel and perpendicular to the fluid flow,
radiation between particle surfaces and the fluid, and lastly,
transfer to or form the environment.   It is impossible at
the present time to obtain either analytical and/or numerical
solutions to the mathematical representation which accounts
for all contributing mechanisms.   For the case where radiation is negligible, Amundsen [1] has presented mathematical rep-
resentations for various combinations of the other mechanisms
with and without the inclusion of heat generation within the
particles.   He obtained analytical solutions by applying
various combinations of Bessel, Fourier and Laplace transforms
to the two coupled partial differential equations.   The
ensuing solutions were found as infinite series of Bessel
functions, trigonometrical functions and a special function,
which had only been tabulated for a narrow parameter range.
The solutions are only valid for the initial and boundary
conditions used in the mathematical representation.

Numerical solutions have been proposed for many of the
combination of mechanisms, and these can be used for any
initial and boundary conditions provided the solution has been
shown to be convergent, stable and compatible.   Thus a single
numerical solution can be applied to any number of physical
situations by simply changing the initial and boundary conditions.   However the assumptions of the mathematical representation must be consistent with the physical system, and of
course, the fewer assumptions employed, the more complex is
the numerical solution.   Often the designer must compromise

between the degree of sophistication of the mathematical model and the computer time required to achieve the solution. Thus any numerical solution must be developed so that the least amount of time is required for the solution, as well as ensuring that the minimum computer storage is used. Although present day main frame computers have considerable memory, and even, home computer memories can be expanded to at least 128K, the numerical analyst must produce a solution which is quick, requires minimum memory and is accurate.

The discourse, which follows, presents numerical solutions for several cases of combinations of heat transfer mechanisms in packed beds and shows how the solutions have been developed, so that the most complex one so far solved numerically, namely; fluid-solid convection, fluid axial conduction and solid internal conduction; meets the requirements of speed, accuracy and minimum memory usage.

2.   THE SIMPLEST MODEL: FLUID-TO-SOLID CONVECTION.

In many particulate packed beds, when the conductivity of the solid material is relatively high, and the system is adiabatic with respect to the environment, then the controlling mechanism is solely convective.   This representation is the simplest and has been used to analyse and design many engineering systems: thermal regenerator heat exchangers, chemical reactors,thermal storage units and transient performance of whole plants, especially during start-up and shut-down.

The two coupled partial differential equations describing this system are obtained by considering heat balances over an increment, dx, of the packed bed, shown in Figure 1, for each phase:

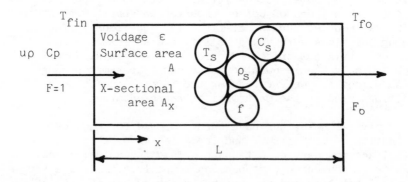

FIGURE 1.   Diagrammatic Representation of a Packed Bed.

Fluid phase

$$\dot{m}Cp \frac{\partial T_f}{\partial x} + \frac{\epsilon}{u} \dot{m}Cp \frac{\partial T_f}{\partial t} = - \frac{\alpha A}{L}(T_f - T_s) \qquad (1)$$

Solid phase

$$(1 - \epsilon) \rho_s C_s A_x \frac{\partial T_s}{\partial t} = \frac{\alpha A}{L}(T_f - T_s) \qquad (2)$$

The assumptions inherent in the derivation of equations (1) and (2) are as follows:

1. The physical properties are constant,
2. The heat transfer coefficient, α, is constant,
3. The fluid is in plug flow,
4. The walls of the packed bed are adiabatic,
5. Axial conduction of heat in either (a) the fluid or (b) the solid phase is negligible, and
6. No thermal gradients exist within the solid phase normal to the transfer surface.

The fluid and solid temperatures in the system will be assumed to have equilibrated prior to the introduction of a step change in the inlet fluid temperature. Thus the initial and boundary conditions are as follows:

$$T_f = T_s = T_i \text{ for } t < 0 \text{ and } 0 \leqslant x \leqslant L \qquad (3)$$

and

$$T_f = T_{fin} \text{ at } x = 0 \text{ for } t \geqslant 0 \qquad (4)$$

The system of equations (1) and (4) are made dimensionless by introducing the following transformations and normalised temperatures,

$$y = \alpha A x / L \dot{m} Cp, \text{ dimensionless distance,} \qquad (5)$$

$$z = \alpha A (t - \epsilon x/u)/(1 - \epsilon) \rho_s C_s A_x L, \text{ dimensionless} \qquad (6)$$
$$\text{time,}$$

$$Y = \alpha A / \dot{m} Cp, \text{ dimensionless bed length,} \qquad (7)$$

$$F = (T_f - T_i)/(T_{fin} - T_i), \text{ normalised fluid} \qquad (8)$$
$$\text{temperature,}$$

and
$$f = (T_s - T_i)/(T_{fin} - T_i), \text{ normalised solid} \qquad (9)$$
$$\text{temperature,}$$

The equations (1) to (4) now become,

$$\frac{\partial F}{\partial y} = - (F - f) \tag{10}$$

$$\frac{\partial f}{\partial z} = (F - f) \tag{11}$$

$F = f = 0$ for $z < 0$ and $0 \leqslant y \leqslant Y$ (12)

and $F = 1$ for $z \geqslant 0$ and $y = 0$ (13)

Schumann [2] obtained the following analytical solution for equations (19) to (13) in terms of the outlet fluid temperature:

$$F_0 = \exp(-Y-z) \sum_{n=0}^{\infty} (z/Y)^{n/2} I_n (2 \sqrt{Yz}) \tag{14}$$

where I_n is a modified Bessels function of the first kind.

A numerical solution for the same set of equations can be found by replacing the derivatives in equations (10) and (11) by finite difference approximations and representing the dimensionless distance and time by a grid as shown in Figure 2.

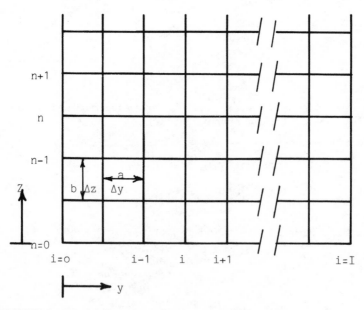

FIGURE 2.   Grid Representation for the Numerical Solution.

Each grid point represents a position in the system at some particular point in time, where a(=Δy) and b(=Δz) are the increment sizes, and i and n denote the length and time positions. The packed bed entrance and exit points are at i=0 and i=I, n=0 is the zero time position, so that Δy = Y/I and I is the total number of length increments. However, the question now arises as to which finite difference approximation to use. Forward-and-backward-difference formulae have leading errors of O(h), where h is the increment size, whilst the central-difference formula has a leading error of O(h^2). Thus, it is possible to use any of these formulae to approximate the derivatives in equations (10) and (11), and of course, to use different ones for each derivative. Cockcroft [3] investigated the stability of all possible combinations of the three finite difference formulae by the Fourier series method [4,5], and the results are shown in Table 1.

Finite difference formula		Stability Condition
Time derivative	Length derivative	
Forward	Forward	Conditionally stable
Forward	Backward	" "
Forward	Central	" "
Backward	Forward	Unconditionally stable
Backward	Backward	" "
Backward	Central	" "
Central	Forward	" "
Central	Backward	" "
Central	Central	" "

TABLE 1   Summary of stability investigation.

All the numerical solutions using forward-difference approximations for the time derivative were found to be conditionally stable, whereas for all other combinations the solutions were unconditionally stable for all values of the time and length increments.

Price [6] used the central-difference formula for both derivatives in obtaining a numerical solution for the set of equations (10) to (13), and the same technique has been used by various workers subject to a variety of initial and boundary conditions [7,8]. Cockcroft [3] found however, that for the non-linear set of equations representing a transient adiabatic chemical reactor, that computationally it was more convenient to employ the backward-difference formula for the time derivative and central-difference formula for the length derivative.

2.1 Numerical solution using the central-difference formula for both derivatives.

The use of the central-difference formula to approximate the derivatives in equations (10) and (11) leads at first sight to an implicit solution, but by judicious manipulation of the resulting algebraic equations leads to an explicit marching solution along each time interval.

The fluid euqation (10) is approximated about the grid point $(n, i+1/2)$ and the following algebraic equation is obtained:

$$\frac{F(n,i) - F(n,i-1)}{a} = -\frac{1}{2} \left\{ \begin{array}{l} F(n,i) + F(n,i-1) - f(n,i) \\ \qquad - f(n,i-1) \end{array} \right\} \qquad (15)$$

whereas the solid equation (11) is approximated about the grid point $(n-1/2, i)$ to give

$$\frac{f(n,i) - f(n-1,i)}{b} = \frac{1}{2} \left\{ \begin{array}{l} F(n,i) + F(n-1,i) \\ - f(n,i) - f(n-1,i) \end{array} \right\} \qquad (16)$$

Equations (15) and (16) are rearranged to provide the fluid and solid temperatures at grid point (n,i) in terms of the other values, i.e.:

$$F(n,i) = a1\, F(n,i-1) + a2\, (f(n,i) + f(n,i-1)) \qquad (17)$$

and $f(n,i) = b1\, f(n-1,i) + b2\, (F(n-1,i) + F(n,i)) \qquad (18)$

where $a1 = (2 - a)/(2 + a) \qquad (19)$

$a2 = a/(2 + a) \qquad (20)$

$b1 = (2 - b)/(2 + b) \qquad (21)$

and $b2 = b/(2 + b) \qquad (22)$

Equation (17) is substituted into (18) to give

$$f(n,i) = c1\, f(n-1,i) + c2\, F(n-1,i) \\ + c3\, f(n,i-1) + c4\, F(n,i-1) \qquad (23)$$

where $c1 = b1/(1 - a2\, b2) \qquad (24)$

$c2 = b2/(1 - a2\, b2) \qquad (25)$

$c3 = a2\, c2 \qquad (26)$

$c4 = a1\, c2 \qquad (27)$

The solution for the initial and boundary conditions (12)

and (13) is as follows:

At n=0, the initial time step, then f is everywhere zero and the fluid temperature distribution is found by solving equation (17) along the grid with F(0,0) = 1.0.   For all other values of n, at i=0, the initial length step, then F(n,0) = 1.0 and f(n,0) is found from

$$f(n,0) = b1\ f(n-1,0) + 2.0\ b2 \qquad\qquad (28)$$

which is simply equation (18) rearranged for the boundary condition (13).   The solid temperature at each grid point is obtained from equation (23) and the corresponding fluid temperature from equation (17).

The computer memory requirements can be minimised in many situations, because either the outlet fluid temperature value is the desired information; or the fluid and solid temperatures along the system at a particular time increment are required.   In these cases, the solid and fluid temperatures at the previous time interval can be overwritten and the storage requirement is simply a two-dimensional array of size (2,I+1).

As mentioned earlier, the numerical solution is unconditionally stable but convergence to three decimal places is only achieved if a = b = 0.1 for Y > 1.0, and for Y < 1.0, then I=10 and again, a = b.

Table 2 lists the dimensionless time for the outlet fluid temperature to reach five values (0.1,0.2,0.5,0.8 and 0.9) at three values of the dimensionless length for various length and time increments.   Convergence is obtained, if a=b=0.1.

	$F_O=$	0.1	0.2	0.5	0.8	0.9
	a=b	z	z	z	z	z
Y=2	0.50	no value	0.259	1.470	3.405	4.712
	0.30	no value	0.246	1.470	3.417	4.722
	0.10	no value	0.240	1.470	3.420	4.728
	0.05	no value	0.240	1.469	3.420	4.728
	0.03	no value	0.240	1.469	3.420	4.728
	0.01	no value	0.240	1.469	3.420	4.728
Y=16	0.90	9.156	11.147	15.499	20.556	23.482
	0.50	9.137	11.132	15.497	20.570	23.503
	0.30	9.132	11.132	15.497	20.575	23.513
	0.10	9.131	11.130	15.497	20.577	23.515
	0.05	9.130	11.130	15.497	20.577	23.515
Y=32	0.90	22.121	25.168	31.499	38.545	42.517
	0.50	22.115	25.151	31.499	38.555	42.526
	0.30	22.111	25.149	31.499	38.559	42.532
	0.10	22.110	25.147	31.499	38.560	42.534

TABLE 2.   Convergence of the numerical solution.

Price [6] showed that the solution was compatible by direct comparison with the analytical solution, equation (14). Table 3 contains the comparison between the numerical and analytical solutions for predicting the outlet fluid temperature at five points (0.1,0.2,0.5,0.8 and 0.9) over a range of dimensionless lengths. The numerical values of the dimensionless time for each point were obtained by linear interpolation over the time interval when the outlet fluid temperature became larger than the required value. The analytical value of the outlet fluid temperature was then calculated from equation (14) using the values of the dimensionless length and the dimensionless time. The agreement over the entire range is within three decimal places.

Y	z	F_O	F'_O
4	0.81665	0.1	0.1000
8	3.27947	0.1	0.1000
16	9.13039	0.1	0.1000
32	22.11008	0.1	0.0999
64	49.93578	0.1	0.1013
2	0.23963	0.2	0.2000
4	1.54299	0.2	0.2000
8	4.53865	0.2	0.2000
16	11.12972	0.2	0.2000
32	25.14727	0.2	0.1999
64	54.35452	0.2	0.1999
1	0.39763	0.5	0.5000
2	1.46941	0.5	0.5000
4	3.48782	0.5	0.5000
8	7.49440	0.5	0.5000
16	15.49730	0.5	0.4999
32	31.49867	0.5	0.4999
64	63.50030	0.5	0.4998
1	1.86008	0.8	0.8000
2	3.42056	0.8	0.7991
4	6.15563	0.8	0.7977
8	11.16599	0.8	0.7989
16	20.57694	0.8	0.7995
32	38.56026	0.8	0.7997
64	73.35670	0.8	0.7995
1	2.90628	0.9	0.8979
2	4.72840	0.9	0.8956
4	7.83617	0.9	0.8975
8	13.36942	0.9	0.8983
16	23.51530	0.9	0.8973
32	42.53644	0.9	0.8983
64	78.79156	0.9	0.8979

F'_O from equation (14) by Schumann

TABLE 3. Comparison of the numerical and analytical solutions.

3. THE AXIAL DISPERSION MODEL

The previous model assumed that conduction in the fluid phase was negligible, however, in packed beds with relatively low mass velocities, there is the likelihood of axial dispersion. This is caused by local trapping of the fluid in the voids, by-passing of voids by the fluid flow, acceleration and deceleration of the fluid through the tortuous channels of the bed, and results in the fluid being dispersed in the axial direction and a degree of backmixing takes place. Thus, this phenomenon is equivalent to an effective axial thermal conductivity, which is at least an order of magnitude greater than the fluid molecular conductivity. A more detailed account of this phenomenon may be found in reference [10].

Hence, the assumption 5(a) of the previous model is relaxed and the two coupled equations describing the axial dispersion model are again obtained by considering heat balances over an increment, dx, of the packed bed for each phase:

Fluid phase

$$\dot{m}Cp \frac{\partial T_f}{\partial x} - k_e A_x \varepsilon \frac{\partial^2 T_f}{\partial x^2} + \frac{\varepsilon}{u} \dot{m}Cp \frac{\partial T_f}{\partial t} = -\frac{\alpha A}{L} (T_f - T_s) \qquad (29)$$

and solid phase

$$(1 - \varepsilon) \rho_s C_s A_x \frac{\partial T_s}{\partial t} = \alpha \frac{A}{L} (T_f - T_s) \qquad (2)$$

The solid phase equation is identical to that used in the simplest model, however the set of equations (2) and (29) is now parabolic due to the second term accounting for the axial dispersion. The variable k_e is the effective axial fluid conductivity and this is related to the effective diffusion coefficient, D_e, as follows:

$$D_e = k_e / \rho_f Cp \qquad (30)$$

Gunn [10] concluded that the axial Peclet number for packed beds, which is defined as follows:

$$Pe_d = \frac{u2R}{\varepsilon D_e} \qquad (31)$$

can be correlated in an acceptable manner by supposing that the total axial dispersion is the sum of convective and diffusive modes of motion. He proposed the following correlation to describe the variance of axial Peclet number with Reynolds number for heat transfer.

$$\frac{1}{Pe_d} = \frac{1}{2.0} + \frac{\epsilon}{\text{╳} RePr} \tag{32}$$

where ╳ is a tortuosity factor equal to 1.5.

The parabolic nature of equation (29) requires two boundary conditions for its solution. The boundary conditions proposed by Danckwerts [11] will be used here to describe the dispersion processes at the entrance and exit to the bed.

These are as follows:

At the entrance

$$T_{fin} = \left[T_f - \frac{k_e A_x \epsilon}{\dot{m}Cp} \quad \frac{\partial T_f}{\partial x} \right]_{x \to 0} \tag{33}$$

and at the exit

$$\left. \frac{\partial T_f}{\partial x} \right|_{x=L} = 0 \tag{34}$$

The equations (2), (29), (33) and (34) must be coupled with an initial condition and a boundary condition relating to the transient nature of the process. The same ones used for the simplest model equations (3) and (4) will be used here. The total set of equations describing this axial dispersion model are made dimensionless by use of the transformations and normalisations (5) to (9), and the following definition:

$$K_e = k_e A_x \epsilon \, \alpha A/L \, (\dot{m}Cp)^2, \text{ the dimensionless conduction parameter} \tag{35}$$

The dimensionless set of equations are as follows:

$$\frac{\partial F}{\partial y} - K_e \frac{\partial^2 F}{\partial y^2} = -(F - f) \tag{36}$$

$$\frac{\partial f}{\partial z} = (F - f) \tag{11}$$

$$F = f = 0 \quad \text{for} \quad z < 0 \quad \text{and} \quad 0 \leqslant y \leqslant Y \tag{12}$$

$$1.0 = \left[F - K_e \frac{\partial F}{\partial y} \right]_{y \to 0} \quad \text{for} \quad z \geqslant 0 \quad \text{at} \quad y \to 0 \tag{37}$$

and $\dfrac{\partial F}{\partial y} = 0$ for $z \geqslant 0$ at $y = 0$ (38)

3.1 Numerical solution using central-difference formulae.

A numerical solution is sought by using central-different formulae for first and second derivatives because the leading error is of $O(h^2)$ for both formulae. The system is again represented by the grid shown in Figure 2.

Equation (35) is approximated about the grid point (n,i) to give the following algebraic equation.

$$\frac{F(n,i+1) - F(n,i-1)}{2a} - \frac{K_e}{a^2}\left\{F(n,i+1) - 2F(n,i) + F(n,i-1)\right\}$$

$$+ F(n,i) - f(n,i) = 0 \qquad (39)$$

Collection of terms in equation (39) provides equation (40), which applies over the range $1 \leqslant i \leqslant I-1$;

$$- U\,F(n,i-1) + V\,F(n,i) - f(n,i) - W\,F(n,I+1) = 0 \qquad (40)$$

where $U = (2K_e + a)/2a^2$ (41)

$V = 2K_e/a^2 + 1$ (42)

$W = (2K_e - a)/2a^2$ (43)

At the bed inlet, $i=0$, equation (40) is combined with a central difference approximation of equation (37) to eliminate the fictitious temperature $F(n,-1)$ and the resulting equation is obtained:

$$GC2\;\;F(n,0) - f(n,0) - GC3\;\;F(n,1) = GC1 \qquad (44)$$

where $GC1 = 2aW/K_e$ (45)

$GC2 = V + GC1$ (46)

$GC3 = W + U$ (47)

At the bed exit, $i=I$, equation (40) is again combined with a central difference approximation of equation (38) to eliminate the fictitious temperature $F(n,I+1)$ and results in the following:

$$-GC3\;\;F(n,I-1) + V\,F(n,I) - f(n,I) = 0 \qquad (48)$$

The solid temperature equation (11) is approximated about the grid point $(n-1/2, i)$ in an identical manner to the previous model and thus equation (18) is obtained again:

$$f(n,i) = b1 \quad f(n-1,i) + b2 \ (F(n-1,i) + \ F(n,i)) \tag{18}$$

Hence at any time interval, n, if the solid and fluid temperatures at the previous interval, n-1, are known, then the fluid and solid temperatures can be found by solving equations (40), (44), (48) and (18). These result in a set of 2 (I+1) equations in 2 (I+1) unknowns and can be solved. The coefficient matrix of this set of equations is pentadiagonal, and the upper and lower diagonals are somewhat sparse. Thus, to avoid excess storage requirements and any problems with zero multiplication, a simpler set of equations is obtained.

Equation (18) is substituted into equations (40), (44) and (48) so that fluid temperatures at time interval n can be found by solving the following set of equations:

At i=0,

$$(GC2-b2) \ F(n,0) - GC3 \ F(n,1) = GC1 + b1 \ f(n-1,0)$$
$$+ \ b2 \ F(n-1,0) \tag{49}$$

for $1 \leqslant i \leqslant I-1$,

$$-U \ F(n,i-1) + (V-b2) \ F(n,i) - W \ F(n,i+1) = bi \ f(n-1,i)$$
$$+ \ b2 \ F(n-1,i) \tag{50}$$

and at i=I,

$$-GC3 \ F(n,I-1) + (V-b2) \ F(n,I) = b1 \ f(n-1,I)$$
$$+ \ b2 \ F(n-1,I) \tag{51}$$

This set of equations has a coefficient matrix, which is tridiagonal, and the inversion is achieved by a special Guass elimination and backward substitution algorithm [4], which accounts for the particular property of the coefficient matrix. The solid temperature profile can be evaluated during the backward substitution by first storing the fluid temperature at the particular point from the previous time interval, then evaluating the new fluid temperature, and solid temperature by equation (18).

The numerical solution is initiated for the initial and boundary conditions (12), (37) and (38) by first solving equations (44), (49) and (48) with f (0,i) = 0.0 to give the initial fluid temperature profile along the packed bed. This set of equations again has a tridiagonal coefficient matrix, and the algorithm for the general time interval can be used here.

The computer memory requirements are a working area of (5,I+1), usually arranged as five vectors, and if the solid and fluid temperatures are overwritten at the next time step, then only an additional (2,I+1) array is required.

3.2 Convergence, stability and compatibility.

For stability and convergence, the central diagonal of the tridiagonal coefficient matrix must be dominant [4]. Diagonal dominance was investigated by investigating various combinations of length increment, a, time increment, b, and the dimensionless conduction parameter, K_e. The time increment, b, was found to have an insignificant effect on the dominancy of the central diagonal elements, but the other two variables, a and K_e, did affect the dominancy. Diagonal dominance is ensured provided the following inequalities are satisfied:

$$K_e > 0.1 \qquad a = 0.1 \qquad\qquad (52)$$

and

$$K_e > 0.1 \qquad a > K_e \qquad\qquad (53)$$

These results were obtained by plotting log K_e against log a [12,13]. The same inequalities hold for the numerical solution at the zero time step, when the central diagonal elements do not contain a contribution from the time increment, b.

No analytical solution completely akin to this model can be found in the literature. Amundsen's solution [2] involves other factors and effects, and is very involved, thus the numerical solution was compared with an analytical solution for the initial temperature profile along the packed bed [12]. This analytical solution for equations (36), (37) and (38) at z=0 and with f=0 is as follows:

$$F(y) = C_1 \exp (k_1 y) + \frac{1 - C_1 + k_1 C_1 K_e \exp (k_2 y)}{1 - k_1 K_e} \qquad (54)$$

where

$$C_1 = \cfrac{1}{1 - k_2 K_e - \cfrac{1 - k_1 K_e}{\dfrac{k_2}{k_1} \exp \{(k_2 - k_1) Y\}}} \qquad (55)$$

$$k_1 = \left\{1 + (1 + 4K_e)^{1/2}\right\} / 2K_e \qquad (56)$$

and

$$k_2 = \left\{1 - (1 + 4K_e)^{1/2}\right\} / 2K_e . \qquad (57)$$

It was considered that if the numerical solution at the
zero time step could be shown to be compatible with the analytical solution, equation (54), then the finite difference solution to the complete set of equations (36),(11), (12), (37) and
(38) was also compatible.

The analytical and numerical solutions were compared over
range of dimensionless bed length, Y, values between 2.5 and
50.   Tables 4 and 5 list the fluid temperature profiles for
Y=2.5 and 10 respectively, for various values of dimensionless
conduction parameter, K_e.   The two solutions are in good
agreement.   The largest discrepancies arise when K_e is large,
that is, when the axial dispersion is very prominent.   These
discrepancies never exceed 2% for any value of y, and the numerical solution is considered sufficiently accurate.Results for
larger values of Y are not included because for values greater
than 25 then 75% of the bed is unaffected at the zero time by
the step change at the entrance.

K_e \ y		0.0 F	0.5 F	1.0 F	1.5 F	2.0 F	2.5 F
2.5	A	0.4769	0.3849	0.3153	0.2656	0.2348	0.2235
	N	0.4719	0.3792	0.3094	0.2593	0.2296	0.2192
0.5	A	0.7320	0.5079	0.3522	0.2455	0.1767	0.1488
	N	0.7308	0.5059	0.3500	0.2437	0.1754	0.1477
0.25	A	0.8284	0.5477	0.3613	0.2385	0.1589	0.1223
	N	0.8278	0.5465	0.3602	0.2377	0.1584	0.1219
0.05	A	0.9544	0.5925	0.3669	0.2272	0.1407	0.0917
	N	0.9544	0.5924	0.3668	0.2272	0.1407	0.0917
0.025	A	0.9761	0.5994	0.3672	0.2249	0.1406	0.0850
	N	0.9761	0.5994	0.3673	0.2250	0.1379	0.0871
0.005	A	0.9950	0.6053	0.3673	0.2229	0.1352	0.0826
	N	0.9950	0.6053	0.3673	0.2228	0.1352	0.0830
0.0025	A	0.9975	0.6060	0.3673	0.2226	0.1349	0.0823
	N	0.9975	0.6060	0.3673	0.2226	0.1349	0.0825
0.0005	A	0.9995	0.6067	0.3673	0.2224	0.1346	0.0821
	N	0.9995	0.6066	0.3672	0.2223	0.1346	0.0819
0.00025	A	0.9997	0.6068	0.3674	0.2224	0.1346	0.0821
	N	0.9997	0.6067	0.3673	0.2223	0.1346	0.0816

A = analytical solution.

N = numerical solution

TABLE 4.   Dimensionless fluid temperature profile for
solutions to the axial disperson model with
Danckwerts' boundary conditions.   Y=2.5.

y K_e		0.0 F	2.01 F	4.03 F	6.05 F	8.07 F	10.00 F
10.0	A	0.2707	0.1577	0.0924	0.0557	0.0370	0.0313
	N	0.2716	0.1586	0.0932	0.0564	0.0375	0.0318
2.0	A	0.5000	0.1803	0.0666	0.0243	0.0090	0.0050
	N	0.5003	0.1833	0.0668	0.0244	0.0091	0.0051
1.0	A	0.6180	0.1784	0.0511	0.0146	0.0042	0.0017
	N	0.6182	0.1786	0.0513	0.0147	0.0042	0.0017
0.2	A	0.8541	0.1534	0.0273	0.0048	0.0008	0.0001
	N	0.8540	0.1534	0.0273	0.0048	0.0008	0.0001
0.1	A	0.9160	0.1452	0.0228	0.0035	0.0005	0.0000
	N	0.9160	0.1453	0.0228	0.0035	0.0005	0.0001
0.02	A	0.9807	0.1365	0.0188	0.0025	0.0003	0.0000
	N	0.9807	0.1365	0.0188	0.0025	0.0003	0.0000
0.01	A	0.9901	0.1353	0.0183	0.0024	0.0003	0.0000
	N	0.9901	0.1353	0.0183	0.0024	0.0003	0.0000
0.002	A	0.9980	0.1342	0.0178	0.0023	0.0003	0.0000
	N	0.9980	0.1342	0.0178	0.0023	0.0003	0.0000
0.001	A	0.9990	0.1341	0.0178	0.0023	0.0003	0.0000
	N	0.9990	0.1341	0.0178	0.0023	0.0003	0.0000

A = analytical solution.

N = numerical solution.

TABLE 5. Dimensionless fluid temperature profiles for solutions to the axial dispersion model with Danckwerts' boundary conditions. Y=10.

The step sizes for the length increment used in the numerical solution gave a stable solution, because the inequalities (52) and (53) were always satisfied. For example, a=0.0025 for Y=2.5 and a=0.025 for Y=25. For the general solution after the evaluation of the initial fluid temperature profile, a time increment of 0.025 provided convergence.

4. THE INTRACONDUCTION MODEL FOR SPHERICAL PARTICLES.

In packed beds with relatively high mass velocities then the assumption of no thermal gradients in the packings in a direction normal to the surface is often invalid. In these cases, the effects of axial dispersion become insignificant and the overall transfer of heat is a combination of convection and conduction within the packing. Thus relaxing of assumption 6 of the simple model in Section 2 must be considered, and the actual geometry of the packing material must be specified.

The two coupled partial differential equations describing

314

this latest system are again obtained by considering heat balances over an increment, dx, of the packed bed, shown in Figure 1, for each phase and assuming the packing shape is spherical:

Fluid phase:

$$\dot{m}Cp \frac{\partial T_f}{\partial x} + \frac{\varepsilon}{u} \dot{m}Cp \frac{\partial T_f}{\partial t} = - \alpha\frac{A}{L} (T_f - T_{ss}) \tag{58}$$

This equation only differs from equation (1) in that the solid temperature, T_{ss}, is that at the spherical particle surface.

Solid phase: For each particle, the thermal behaviour is described by,

$$\rho_s C_s \frac{\partial T_s}{\partial t} = k_s \left(\frac{\partial^2 T_s}{\partial r^2} + \frac{2}{r} \frac{\partial T_s}{\partial r} \right) \tag{59}$$

The equations (58) and (59) are coupled by the heat balance at the fluid-solid interface, that is at r=R:

$$-k_s \frac{\partial T_s}{\partial r} = \alpha (T_{ss} - T_f) \tag{60}$$

and the system of equations is completed by the symmetry boundary condition at the particle centre, at r=0,

$$\frac{\partial T_s}{\partial r} = 0 \tag{61}$$

The initial and boundary conditions will again be taken for an equilibrated packed bed system prior to a step change in the inlet fluid temperature:

$$T_f = T_s = T_i \quad \text{for} \quad t < 0, \ 0 \leqslant r \leqslant R \quad \text{and} \quad 0 \leqslant x \leqslant L \tag{62}$$

$$\text{and} \quad T_f = T_{fin} \quad \text{at} \quad x=0 \quad \text{for} \quad t \geqslant 0. \tag{4}$$

The system of equations (58), (59), (60) and (61), and the initial and boundary conditions (62) and (4) are made dimensionless by introducing the transformations (5), (6), (7), and normalised temperatures (8) and (9), and the following transformations:

$$s = r/R, \text{ normalised radius} \tag{63}$$

and $\quad$ Bi $= \quad \alpha R/k_s$, Biot number $\hfill$ (64)

The intrasphere conduction model in dimensionless form is as follows:

$$\frac{\partial F}{\partial y} = - (F - f_s) \hfill (65)$$

where f_s is the dimensionless surface temperature of a particle

$$3 \, Bi \frac{\partial f}{\partial z} = \frac{\partial^2 f}{\partial s^2} + \frac{2}{s} \frac{\partial f}{\partial s} \hfill (66)$$

with the following boundary conditions:

$\quad$ F = 1 $\;$ for $\;$ z $\geqslant$ 0 $\quad$ and $\quad$ y = 0 $\hfill$ (13)

$$\frac{\partial f}{\partial s} = - Bi(F - f) \text{ at } s=1, \text{ for } z \geqslant 0 \text{ and } 0 \leqslant y \leqslant Y \hfill (67)$$

and $\dfrac{\partial f}{\partial s} = 0 \qquad$ at s=0, for z $\geqslant$ 0 and 0 $\leqslant$ y $\leqslant$ Y $\hfill$ (68)

with the initial condition,

$\quad$ F = f = 0 $\;$ for $\;$ z < 0, $\;$ 0 $\leqslant$ s $\leqslant$ 1 $\;$ and $\;$ 0 $\leqslant$ y $\leqslant$ Y $\hfill$ (69)

$\qquad$ The two coupled partial differential equations (65) and (66) are a set of parabolic equations. Equation (65) itself is hyperbolic, but the set is dominated by the parabolic nature of equation (66).

$\qquad$ Rosen [14] obtained an analytical solution for the analogous physical situation of mass transfer to spherical particles taking into account diffusion within the solid. Handley and Heggs [15] presented a numerical solution which used the Crank-Nicolson approximation [4] for equation (66) and central-difference formula for equation (65). The grid representation for this numerical solution is now three-dimensional and this is shown in Figure 3 where c(=Δs), j denotes a grid position in the particle dimension, and j=0 and j=J are the points corresponding to the centre and surface of the spherical particle respectively. All other symbols in Figure 3 are identical to those shown in Figure 2.

$\qquad$ The same numerical approach has been used for other initial and boundary conditions for spherical particles and other particle geometries: plates, solid cylinders and hollow cylinders [16,17,18,19].

316

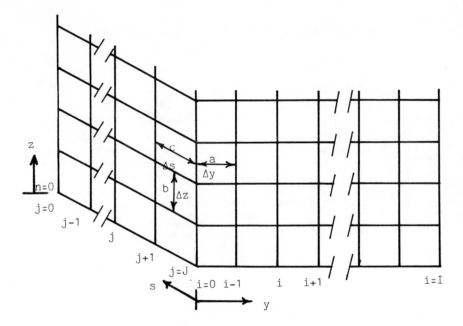

FIGURE 3   Grid representation for three dimensional system.

4.1 Numerical solution using central-difference and Crank-Nicolson approximations.

The fluid equation (65) is again approximated by central-difference formula about the grid point $(n, i+1/2)$ and the fluid temperature at grid point (n, i) is given by

$$F(n,i) - a2\ f(n,i,J) = a1\ F(n,i-1) + a2\ f(n,i-1,J) \qquad (70)$$

Note that equation (70) is identical to equation (17) except that the solid temperatures are those at the surface of the spherical particle.

The solid equation (66) is represented by the Crank-Nicolson 6-point implicit form, because this representation is unconditionally stable [4].   The approximation is applied at grid point $(n-1/2, i, j)$ to give

$$-P_j\ f(n,i,j+1) + (1+M)\ f(n,i,j) - Q_j\ f(n,i,j-1) =$$

$$P_j\ f(n-1,i,j+1) + (1-M)\ f(n-1,i,j) + Q_j\ f(n-1,i,j-1) \qquad (71)$$

where $M = b/(3\ Bi\ c^2)$ \qquad (72)

$$P_j = M(1+1/j)/2 \tag{73}$$

and $\quad Q_j = M(1-1/j)/2 \tag{74}$

Equation (71) applies over the range $1 < j < J-1$.

At the sphere centre, L'Hopital's rule is first applied to equation (66) to give

$$Bi \; \frac{\partial f}{\partial z} = \frac{\partial^2 f}{\partial s^2} \tag{75}$$

and then the Crank-Nicolson approximation is applied to equation (75) and combined with a central-difference approximation to equation (68) to give at $j=0$:

$- PC1 \; f(n,i,1) + PC2 \; f(n,i,0) = PC1 \; f(n-1,i,1)$

$+ PC3 \; f(n-1,i,0) \tag{76}$

where $PC1 = 3M \tag{77}$

$\quad PC2 = 1 + 3M \tag{78}$

and $\quad PC3 = 1 - 3M \tag{79}$

At the particle surface, $j=J$, equation (71) is combined with a central-difference approximation to equation (67) to give:

$- PS1 \; F(n,i) + PS2 \; f(n,i,J) - M \; f(n,i,J-1) =$

$\quad PS1 \; F(n-1,i) + PS3 \; f(n-1,i,J) + M \; f(n-1,i,J-1) \tag{80}$

where $PS1 = c \; Bi \; M \; (1+1/J) \tag{81}$

$\quad PS2 = 1 + M + PS1 \tag{82}$

and $\quad PS3 = 1 - M - PS1 \tag{83}$

Thus at any time interval, n, and length position, i, the fluid and solid temperatures can be found by solving equations (70), (76), (71) and (80) provided the temperatures at the previous length position $(n,i-1)$ and time position $(n-1,i)$ are known. These $(J+2)$ equations have a coefficient matrix which is tridiagonal, and so the algorithm used for the numerical solution of the axial dispersion model can also be used here.

The solution is started by solving equation (70) at $n=0$ with f everywhere zero, and at $i=0$, for all other values of n, the solid profile at the bed entrance is obtained by solving equations (76), (71) and (80) with $F=1.0$. This is a set of

(J+1) equations, which also has a tridiagonal coefficient matrix and so the special algorithm can be used again.

The working storage required by the algorithm is (5,J+1), usually arranged as five vectors, and if the solid and fluid temperatures are overwritten at the next time step, then an additional (J+2,I+1) array is required.

4.2 Convergence, stability and compatibility.

The stability of the hyperbolic and parabolic equations is known, but the manner of coupling these two equations by the flux boundary conditions at the particle surface, equation (67) may seriously affect the overall stability of the numerical analysis. Parker and Crank [20], and Keast and Mitchell [21] have discussed the stability of the Crank-Nicolson formula for various parabolic partial differential equations and boundary conditions, and found that persistent discretization errors may occur in the solution. These errors are dependent upon the increment size used in the approximation of the parabolic equation.

The numerical solution was initially checked for stability for the zero length condition, that is equations (66), (67), (68) and (13). Table 6 lists the values of f_s, particle

			z=0.1	2.0	F_s 0.999	
	b	c	F_s	F_s	Steps	F_s
Bi=0.01	0.01	0.20	0.09659	0.86440	693	0.99901
	0.05	0.20	0.09511	0.86420	139	0.99905
	0.10	0.20	0.09875	0.86365	69	0.99930
	0.10	0.10	0.09870	0.86352	69	0.99950
	0.10	0.05	0.09880	0.86349	69	0.99958
	0.20	0.20	0.69813	0.86367	35	0.99926
Bi= 0.1	0.01	0.20	0.20179	0.84387	870	0.99937
	0.05	0.20	0.20453	0.84391	174	0.99937
	0.10	0.20	0.22435	0.84402	87	0.99937
	0.10	0.10	0.24660	0.84372	87	0.99937
	0.10	0.05	0.25536	0.84085	87	0.99938
	0.20	0.20	0.64813	0.84433	44	0.99943
Bi=10.0	0.01	0.20	0.30626	0.86740	2875	0.99990
	0.05	0.20	0.30776	0.86741	575	0.99990
	0.10	0.20	0.31267	0.86745	288	0.99990
	0.10	0.10	0.44365	0.86285	284	0.99990
	0.10	0.05	0.53995	0.86182	283	0.99990
	0.20	0.20	0.79603	0.86760	144	0.99990

TABLE 6   Stability of the zero length condition.

surface temperature, for z=0.1 and 2.0, and the number of time steps for the value of f_s to reach 0.999 and the value at that time, for b=0.01, 0.05, 0.1 and 0.2 and for c=0.1 and 0.05.

Decreasing the size of b, the time step, only affects the value of f_s at very small values of z, which is caused by the discontinuity due to the initial condition, although these effects are damped away when z has reached 2.0. The number of increments within the particle has little effect on the values of f_s until the value of Bi has increased to 10.0. This value of Bi easily represents the upper limit one would expect for a physical situation, and the solution is stable and convergent for values less than 10.0 and shows no effect of any discretization errors.

The mumerical solution has been checked for compatibility by comparison with the infinite integral solution derived by Rosen [14] for the case of solid diffusion in fixed beds. Table 7 shows the comparison between Rosen's results and the numerical solution. Excellent agreement between the results indicates that the numerical solution is compatible with the original partial differential equations.

5. THE AXIAL DISPERSION AND INTRACONDUCTION MODEL FOR SPHERICAL PARTICLES.

In many situations, a three mechanistic representation of the heat transfer is closer to the physical situation; that is fluid-to-particle convection, internal solid conduction and fluid axial conduction. These mechanisms have been shown to occur in packed bed experiments [22], which have been used to predict lumped heat transfer coefficients, α , for use in equations (1) and (2). Thus in this model both assumptions 5(a) and 6 of the simple model are relaxed.

The two coupled partial differential equations describing this system are again obtained by considering heat balances over an increment, dx, of the packed bed, shown in Figure 1, for each phase and assuming the packing is spherical:

Fluid phase:

$$\dot{m}Cp \frac{\partial T_f}{\partial x} - k_e A_x \varepsilon \frac{\partial^2 T_f}{\partial x^2} + \frac{\varepsilon}{u} \dot{m}Cp \frac{\partial T_f}{\partial t} = -\alpha \frac{A}{L} (T_f - T_{ss}) \quad (84)$$

This equations only differs from equation (29) in that the solid temperature, T_{ss}, is that at the spherical particle surface.

Y=5; Bi=0.125

F_o	z'	z
0.049	0.75	0.75
0.117	1.50	1.51
0.252	2.66	2.63
0.454	4.13	4.13
0.675	6.00	6.00
0.849	8.25	8.25
0.992	15.00	15.01

Y=40; Bi=1.0

F_o	z'	z
0.015	21.00	21.08
0.152	30.00	30.00
0.361	36.00	36.00
0.484	39.00	39.00
0.604	42.00	42.01
0.799	48.00	48.00
0.970	60.00	59.98

Y=10; Bi=20.0

F_o	z'	z
0.128	1.50	1.56
0.263	3.00	3.00
0.421	5.25	5.22
0.517	7.05	7.02
0.654	10.50	10.47
0.778	15.00	15.16
0.948	30.00	29.89

z' was calculated from Rosen's results,

z   was computed using the intraparticle model
    employing the following step sizes;

    a = b = 0.2; c = 0.1;

TABLE 7   Comparison of Rosen's & the intraparticle
          model resuts.

Solid phase:

$$\rho_s C_s \frac{\partial T_s}{\partial t} = k_s \left(\frac{\partial^2 T_s}{\partial r^2} + \frac{2}{r} \frac{\partial T_s}{\partial r} \right) \tag{59}$$

Both equations (84) and (59) are parabolic in nature and require two boundary conditions each. For equation (84) the Danckwerts [11] boundary conditions will be employed; in a similar manner to the axial dispersion model in Section 3.

At the entrance:

$$T_{fin} = \left[T_f - \frac{k_e A_x \varepsilon}{\dot{m} Cp} \frac{\partial T_f}{\partial x} \right]_{x \to 0} \tag{33}$$

and at the exit

$$\frac{\partial T_f}{\partial x} \bigg|_{x=L} = 0 \tag{34}$$

The boundary condition for equation (59) at the particle surface couples equations (84) and (59), at r=R:

$$- k_s \frac{\partial T_s}{\partial r} = \alpha (T_{ss} - T_f) \tag{60}$$

and symmetry is again assumed at the particle centre, at r=0,

$$\frac{\partial T_s}{\partial r} = 0 \tag{61}$$

The initial and boundary conditions for the set of equations will again be for an equilibrated packed bed system prior to a step change in the inlet fluid temperature,

$$T_f = T_s = T_i \quad \text{for} \quad t < 0, \ 0 \leqslant r \leqslant R \text{ and } 0 \leqslant x \leqslant L \tag{62}$$

and $T_f = T_{fin}$ at $x < 0$ for $t > 0$. $\tag{85}$

The system of equations (84), (59), (33), (34), (60), (61), (62) and (85) are made dimensionless by the transformations (5), (6), (7), (35), (63) and (64) and normalised temperatures (8) and (9) to give the following:

$$K_e \frac{\partial^2 F}{\partial y^2} - \frac{\partial F}{\partial y} = (F - f_s) \tag{86}$$

$$3Bi \ \frac{\partial F}{\partial z} = \frac{\partial^2 f}{\partial s^2} + \frac{2}{s} \ \frac{\partial f}{\partial s} \tag{66}$$

$$1.0 = \left[F - K_e \ \frac{\partial F}{\partial y} \right]_{y \to 0} \qquad \text{for} \quad z \geqslant 0 \text{ at } y \to 0 \tag{37}$$

$$\frac{\partial F}{\partial y} = 0 \quad \text{for} \quad z \geqslant 0 \text{ at } y = Y \tag{38}$$

$$\frac{\partial f}{\partial s} = -Bi \ (F-f) \text{ at } s=1, \text{ for } z \geqslant 0 \text{ and } 0 \leqslant y \leqslant Y \tag{67}$$

$$\frac{\partial f}{\partial s} = 0 \text{ at } s=0, \text{ for } z \geqslant 0 \text{ and } 0 \leqslant y \leqslant Y \tag{68}$$

$$F = f = 0 \quad \text{for} \quad z < 0, \ 0 \leqslant s \leqslant 1 \text{ and } 0 \leqslant y \leqslant Y \tag{69}$$

and initial condition

$$F = 1.0 \quad \text{at} \quad x < 0 \quad \text{for} \quad z \geqslant 0 \tag{87}$$

Analytical solutions for this representation have been obtained for a number of inlet responses, and the Laplace transform approach used to generate the solution in the imaginary plane. These solutions were then used to obtain first, second and third moments of the outlet response, F_o, for comparison of the contribution of the three mechanisms to the overall transfer, [22,23,24]. Saez and McCoy [25] have recently presented a similar treatment for the same representation, and obtained solutions of the outlet fluid temperature response in terms of either Hermite or Laguerre polynominal expansions for an impulse input temperature. However they state that when the initial boundary condition is not uniform then the solution method cannot be used.

Heggs and O'Sullivan [26] proposed a numerical solution for this representation which is essentially a combination of the numerical schemes for the axial dispersion model and for the intraconduction model for spherical particles included in Sections 3.1 and 4.1 respectively. The system is represented by a three-dimensional grid as shown in Figure 3, and central-difference formulae are used to approximate equations (86) and the Crank-Nicolson formula for equation (66).

5.1 Numerical solution using central-difference and Crank-Nicolson approximations.

The application of central-difference formulae to equation (86) and its associated boundary conditions (37) and (38) is identical to that previously developed in Section 3.1 and so only the final rearranged algebraic equations will be given here.

At the bed entrance, i=0:

$$GC2\ F(n,0) - f(n,0,J) - GC3\ F(n,1) = GC1 \tag{88}$$

over the range, $1 \leqslant i \leqslant I-1$:

$$- U\ F(n,i-1) + V\ F(n,i) - f(n,i,J) - W\ F(n,i+1) = 0 \tag{89}$$

and at the bed exit, i=I:

$$- GC3\ F(n,I-1) + V\ F(n,I) - f(n,I,J) = 0 \tag{90}$$

Note: equations (88), (89) and (90) only differ to equations (44), (40) and (48) in that the solid temperature is the surface value.

The algebraic equations representing the Crank-Nicolson approximation to equation (66) are identical to those obtained in Section 4.1, and for completeness are again repeated here:

At the particle centre, j=0:

$$- PC1\ f(n,i,1) + PC2\ f(n,i,0) = PC1\ f(n-1,i,1) +$$

$$PC3\ f(n-1,i,0) \tag{76}$$

over the range, $1 \leqslant j \leqslant J-1$:

$$- P_j\ f(n,i,j+1) + (1+M)\ f(n,i,j) - Q_j\ f(n,i,j-1) =$$

$$P_j\ f(n-1,i,j+1) + (1-M)\ f(n-1,i,j) + Q_j\ f(n-1,i,j-1) \tag{71}$$

and at the particle surface, j=J:

$$- PS1\ F(n,i) + PS2\ f(n,i,J) - M\ f(n,i,J-1) =$$

$$PS1\ F(n-1,i) + PS3\ f(n-1,i,J) + M\ f(n-1,i,J-1) \tag{80}$$

Hence for $n > 0$, the algebraic equations (88), (89), (90), (76), (71) and (80) describe the transfer of heat, and if all temperatures at (n-1) are known, then there are (I+1) (J+2) equations and unknowns. The equations are arranged in matrix form:

$$\underline{A1}\ .\ \underline{w} = \underline{B1} \tag{91}$$

where $\underline{A1}$ is the coefficient matrix, $\underline{w}$ is the vector of unknown temperatures at n, and $\underline{B1}$ is the vector of known temperatures at (n-1). The equations are judiciously arranged so that $\underline{A1}$ is a pentadiagonal matrix, as shown in Figure 4, of size (I+1) (J+2). The fluid and solid temperature equations at each axial position have been grouped sequentially to give this particular arrangement. The sparse lower and upper diagonals reflect the fluid axial conduction equations (88), (89) and (90), whilst the inner tridiagonals represent the particle conduction equations (76), (71) and (80).

At the zero time step, n = 0, the particle temperatures are zero, see initial condition equation (69), and the initial fluid temperature profile is obtained from equations (88), (89) and (90). The coefficient matrix for these equations is tridiagonal, and the inversion is achieved by a special Gauss elimination and backward substitution algorithm [4], which accounts for the particular property of the coefficient matrix.

5.2 Algorithm for the solution of A1. w = B1

The algorithm accommodates the pentadiagonal property of the coefficient matrix A1. The elements of the matrix can be stored in three vectors by storing the sparse lower and upper pentadiagonal elements in the lower and upper tridiagonal zeros. However, to illustrate the algorithm, we will consider the matrix A1. During the backward elimination, the non-zero elements of the upper pentadiagonal and tridiagonal are reduced to zero and the resubstitution accounts for the manipulated lower pentadiagonal and tridiagonal elements.

The calculation for the backward elimination is contained in vectors β and γ, and the following evaluation of the elements of the two vectors is shown in terms of element positions in matrix A1 and vector B1.

5.2(a) Backward Elimination

Vectors β and γ contain $(I + 1)(J + 2)$ elements

For i = (I + 1), I, ------, 1, 0:

$$\beta(k) = A1(k,k) \text{ and } \gamma(k) = B1(k,k) \text{ for } k=i \ (J + 2) \qquad (92)$$

$$\beta(k) = A1(k,k) - A1(k,k+1)A1(k+1,k)/\beta \ (k+1)$$

and $\quad \gamma(k) = B1(k,k) - A1(k,k+1) \ \gamma(k+1)/\beta \ (k+1)$

for $\quad$ k=i(J+2)-j and j = 1, 2,----, J-1, J. $\qquad$ (93)

$$\beta(k) = A1(k,k) - A1(k+J+2,k)A1(k,k+j+2)/\beta \ (k+J+2)$$

$$- A1(k,k+1)A1(k+1,k)/ \ \beta(k+1)$$

and $\quad \gamma(k) = B1(k,k) - A1(k,k+J+2)\gamma \ (k+J+2)/ \ \beta(k+J+2)$

$$- A1(k,k+1) \ \gamma(k+1) / \ \beta(k+1)$$

for $\quad$ k = i(J+2)+1 and i $\neq$ I+1 $\qquad$ (94)

Equations (92) are for the centre of the particle, equations (93) for the interior and surface points of each particle and equations (94) are for the fluid. The second terms on the left hand side of the equality sign in equations (94) account for the reduction of the non-zero elements of the upper pentadiagonal.

$$
\begin{bmatrix}
GC2 & -1 & & & & -GC3 & & & & & \\
-PS1 & PS2 & -M & & 0 & & & & & & \\
& -P_j\ PT1 & -Q_j & & & & & & & & \\
& & -PC1 & PC2 & 0 & & & & & & \\
& & & -P_j\ PT1 & -Q_j & 0 & & & & & \\
& & & & -PC1 & PC2 & 0 & & & & \\
& 0 & V & -1 & & & \cdot & & & & \\
-U & & 0 & -PS1 & PS2 & -M & & \cdot & & & \\
& 0 & & & -P_j\ PT1 & -Q_j & & & \cdot & & \\
& & & & & -CG3 & 0 & & & & \\
& & & & 0 & & V & -1 & & & \\
& & & & & 0 & & -PS1 & PS2 & -M & \\
& & & & & & & & -P_j\ PT1 & -Q_j & \\
& & & & & & & & & -P_j\ PT1 & -Q_j \\
& & & & & & & & 0 & & -PC1\ PC2
\end{bmatrix}
$$

$$
\cdot
\begin{bmatrix}
F(n,0=\\
f(n,0,J)\\
f(n,0,J-1)\\
\cdot\\
f(n,0,1)\\
f(n,0,0)\\
F(n,1)\\
f(n,1,J)\\
f(n,1,j-1)\\
\cdot\\
\cdot\\
\cdot\\
f(n,I-1,1)\\
f(n,I-1,0)\\
F(n,I)\\
f(n,I,J)\\
f(n,I,J-1)\\
\cdot\\
f(n,I,1)\\
f(n,I,0)
\end{bmatrix}
=
\begin{bmatrix}
\text{GC1}\\
\text{Equation (80) } i=0\\
\text{Equation (71) } i=0,\ j=J-1\\
\cdot\\
\text{Equation (71) } i=0,\ j=1\\
\text{Equation (76) } i=0\\
0\\
\text{Equation (80) } i=1\\
\text{Equation (71) } i=1,\ j=J-1\\
\cdot\\
\cdot\\
\cdot\\
\text{Equation (71) } i=I-1,j=1\\
\text{Equation (76) } i=I-1\\
0\\
\text{Equation (80) } i=I\\
\text{Equation (71) } i=I,j=J-1\\
\cdot\\
\text{Equation (71) } i=I,j=1\\
\text{Equation (76) } i=I
\end{bmatrix}
$$

FIGURE 4. Schematic of A1.w = B1. PT1=1+M

5.2(b)   Forward Resubstitution.

The vector $\underline{w}$ contains the $(I+1)(J+2)$ temperatures at n^{th} time step.

For k=1

$$w(k) = \gamma(k)/\beta(k) \qquad (95)$$

For i=0, 1,----,I:

for k=i(J+1) + j and j = 2, 3,---,J + 2

$$w(k) = \left\{\gamma(k) + A1(k,k-1)w(k-1)\right\}/\beta(k) \qquad (96)$$

for k = i(J + 1) + 1 for i ≠ 0

$$w(k) = \left\{\gamma(k) + A1(k,k-J-2)w(k-J-2)\right\}/\beta(k) \qquad (97)$$

Equation (95) gives the fluid temperatures at the entrance, equation (96) gives the fluid at all other points and (97) gives the solid temperatures at each axial point.

Vector $\underline{\beta}$ only needs to be evaluated once, because the co-efficients of matrix $\underline{A1}$ remain constant over all time steps, similarly $A1/\beta$ in equations (93) and (94) for γ need only be evaluated once, and stored in a vector $\underline{\delta}$. Vector γ needs to be evaluated for each time step, because the temperatures in most of the elements of $\underline{B1}$ take the values at the previous time step.

5.3   Convergence, stability and compatibility.

To maintain diagonal dominance in $\underline{A1}$ the following in-equalities must be satisfied:

$$a < 2 \quad \text{for } K_e < 0.05 \text{ and } Y < 100 \qquad (98)$$

$$\text{and } a=0.1 \text{ for } K_e > 0.05 \text{ and } Y < 100 \qquad (99)$$

The numerical solution has been investigated over a wide range of increment sizes, and provided the inequalities (98) and (99) are satisfied, then no instabilities were observed. Convergence was checked by reducing the parameters K_e and Bi for fixed values of Y and comparing the outlet response tem-peratures with those obtained from the solution for equations (1) and (2), the simple model. Table 8 lists the dimensionless time values for various values F_o(0.1 to 0.9)for both the simple and this model for two values of Y(=5 and 20). Four combinations of K_e and Bi (=0.1 and 0.05) have been used and the increment sizes employed were a=b=c=0.1. The outlet tem-perature response of this present model is converging to the Schumann as both K_e and Bi are decreased. Although the numerical solution has not been compared with an analytical solution, the results in Table 8 infer compatibility with the simple model in the limit of low values of K_e and Bi.

DIMENSIONLESS TIME z VALUES

$Y=5$ F_o	Simple Model	K_e Bi 0.1 0.1	K_e Bi 0.1 0.05	K_e Bi 0.05 0.1	K_e Bi 0.05 0.05
0.1	1.377	1.180	1.209	1.245	1.273
0.2	2.256	2.074	2.100	2.136	2.163
0.3	3.014	2.860	2.882	2.913	2.935
0.4	3.742	3.624	3.641	3.668	3.683
0.5	4.491	4.416	4.427	4.444	4.454
0.6	5.304	5.283	5.286	5.293	5.295
0.7	6.247	6.292	6.286	6.280	6.273
0.8	7.445	7.583	7.564	7.538	7.518
0.9	9.275	9.567	9.526	9.468	9.426
$Y=20$					
0.1	12.268	11.801	11.870	11.938	12.057
0.2	14.564	14.214	14.266	14.364	14.417
0.3	16.340	16.091	16.129	16.210	16.248
0.4	17.938	17.787	17.812	17.874	17.899
0.5	19.498	19.449	19.459	19.502	19.512
0.6	21.122	21.185	21.180	21.201	21.195
0.7	22.932	23.124	23.101	23.095	23.072
0.8	25.143	25.500	25.454	25.416	25.369
0.9	28.377	28.988	28.907	28.816	28.735

TABLE 8   Calculated Outlet Temperature Responses a=b=c=0.1

6. ACKNOWLEDGEMENTS.

The author is indebted to his past research students;
C.S.Cockcroft, K.J.Carpenter, A.J.St.J.Main, E.Amooie-Foumeny
and P.A.O'Sullivan, who have all helped develop these numerical
solutions for simulating various physical situations.
Mrs.M.Lawn continues to type my hand written scripts and thus,
deserves praise indeed.

REFERENCES.

1. AMUNDSEN, W.R.,   Solid-fluid interactions in fixed and
    moving beds. Ind. Eng. Chem., Vol. 48, pp.1-50, 1956.

2. SCHUMANN, T.E.W.,   Heat transfer. A fluid liquid flowing
    through a porous prism. J.Franklin Inst., Vol.208,
    pp. 405-416, k929.

3. COCKCROFT, C.S. Invesitgation of a thermally regenerative
    reactor system, Ph.D. Thesis, Leeds University, 1976.

4. SMITH, G.D. Numerical Solution of Partial Differential
    Equations, Oxford University Press, London, 1965.

5. AMES, W.F. Nonlinear Partial Differential Equations in
    Engineering, Academic Press, New York, 1965.

328

6. PRICE, C.B.A.   Heat and momentum transfer in thermal regenerators, Ph.D. Thesis, London University, 1964.

7. WILLMOTT, A.J.   Digital computer simulation of a thermal regenerator, Int.J.Heat Mass Transfer, Vol.7, pp.1291-1302, 1964.

8. HEGGS, P.J. and CARPENTER, K.J.   Effect of fluid hold-up on contraflow regenerators, Trans.Instn.Chem.Engrs., Vol. 54, pp. 232-238, 1976.

9. HEGGS, P.J.   Transfer processes in packings used in thermal regenerators, Ph.D. Thesis, Leeds University, 1967.

10. GUNN, D.J.   Mixing in packed and fluidised beds, The Chemical Engineer, Vol.219, pp.153-172, 1968.

11. DANCKWERTS, P.V.   Continuous Flow Systems, Chem.Eng.Sci., Vol.2. p.1., 1953.

12. MAIN, A.J.St.J.   An experimental investigation into heat transfer and momentum effects in packed beds, Ph.D. Thesis Leeds University, 1978.

13. AMOOIE-FOUMENY, E.   Heat Transfer in cyclic regenerators, Ph.D. Thesis, Leeds University, 1983.

14. ROSEN, J.B.   General numerical solution for solid diffusion in fixed beds.   Ind.Eng.Chem.,Vol.46, p.1590, 1954.

15. HANDLEY, D. and HEGGS, P.J.   The effect of thermal conductivity of the packing material on transient heat transfer in packed beds, Int.J.Heat Mass Transfer, Vol.12 pp. 549-570, 1969.

16. HEGGS, P.J. and CARPENTER, K.J.   Prediction of a dividing line between conduction and convection effects in regenerator design, Trans. Instn. Chem. Engrs., Vol. 56, pp.86-90, 1978.

17. HEGGS, P.J.   Intraplate conduction effects in fixed bed transient heat transfer, Can.J.Chem.Engg., Vol.47, pp. 373-377, 1969.

18. HEGGS, P.J.   Conduction effects in Hasche tiles, Heat Transfer 1970, Cu 3.4, Elsevier Publishing Company, Netherlands, 1970.

19. WILLMOTT, A.J.   The regenerative heat exchanger computer representation, Int.J.Heat Mass Transfer, Vol.12, pp.997-1014, 1969.

20. PARKER, I.B. and CRANK, J.   Persistent discretization errors in partial differential equations, Computer J., Vol. 7, p. 163, 1964.

21. KEAST, P. and MITCHELL, A.R.   On the stability of the Crank-Nicolson formula under derivative boundary conditions, Computer J. Vol. 9, p.110, 1966.

22. JEFFRESON, C.P.   Prediction of breakthrough curves in packed beds, AIChEJ, Vol.18, pp.409-415, 1972.

23. BRADSHAW, A.V. et.al.   Heat transfer between air and nitrogen and packed beds on non-reacting solids, Trans. Inst.Chem.Engrs.Vol.48, pp. 77, 1970.

24. RASMUSSEN, A. and NERETNIEKS,I.   Exact solution of a model for diffusion in particles and longitudinal dispersion in packed beds, AJChEJ, Vol. 26, p.686, 1980.

25. SAEZ, A.E. and McCOY, B.J.   Transient analysis of packed-bed thermal storage system, Int. J. Heat Mass Transfer, Vol. 26, pp.49-54, 1983.

26. HEGGS, P.J. and O'SULLIVAN, P.A.   A numerical solution of the effects of fluid convection and axial conduction and solid internal conduction on the heat transfer in porous media, pp.184-196.   Thermal Problems Volume III, Pineridge Press, Swansea, 1983.

27. O'SULLIVAN, P.A.   The effects of convection and conduction on gas-solid heat transfer, Ph.D. Thesis (in preparation), Leeds University, 1984.

CHAPTER 13

FINITE ELEMENT MODELLING OF THE MOLD-METAL INTERFACE IN
CASTING SIMULATION WITH COINCIDENT NODES OR THIN ELEMENTS

M. SAMONDS, R.W. LEWIS, K. MORGAN and R. SYMBERLIST

Department of Civil Engineering, University College of Swansea,
Swansea, U.K.

ABSTRACT

When finite element analysis is employed to model the
solidification process in metal castings, the problem arises
as to how the critical interface between mold and metal might
be treated. This paper introduces the Coincident Node Tech-
nqiue, and compares it with the use of very thin elements.
Actual castings have been modelled and the numerical results
are compared with the experimental measurements.

INTRODUCTION

Numerical modelling of casting processes holds great
promise for the foundry and has therefore been the object of
considerable attention during the past two decades [1,2,3].
Ultimately, a flexible heat analysis program coupled with a
complete CAD system comprising geometric modelling, pre- and
post-processing graphics, other analysis programs (e.g. stress),
NC machine links and so on, will greatly improve the design of
dies and molds [4,5]. For example, hot spots and areas where
shrinkage porosity are likely can be determined before the first
casting is poured, and steps taken to ensure directional solidification. The placement and size of risers may be optimized.
Thermal gradients and solidification may be modulated to produce the desired mechanical properties in the finished casting.
Thermal stress cracking in the die and casting may be min-
imized. The effect of air or water lines and heaters may be
investigated. There is an enormous amount of work yet to be
done in this field.

The area of present concern is the interface condition
between metal and mold, an absolutely vital factor which in
many cases controls the solidification rate more than any
other parameter. Two problems arise for the numerical modeller

332

in this respect: 1) What sort of physical data to use for a
heat transfer coefficient and 2) how is this data incorporated
into the numerical algorithm. The variation of the heat transfer coefficient is a very complicated phenomenon which depends,
among other things, upon the thickness and material of a die
coat, surface roughness, the geometry, pressure and temperature of the solidfying metal, cooling rate, and the formation
of an air gap. Sun [6], Prates et al. [7], Sully [15] and
Nelson [16], discuss some experimental techniques for determining the value of the coefficient and how it changes with
time. Previous workers using a finite difference spatial discretization have generally treated the interface as a series
resistance to be incorporated into the total resistance
between one node in the mold and the other in the metal [8,9].
An alternative approach was taken by Jeyarajan and Pehlke [10],
who used two slightly separated nodes at the interface between
metal and chill. Rather than conduction, the heat flux
between these nodes was described as $q = h (T_{METAL} - T_{CHILL})$.

A similar idea will be presented here in the finite
element context, wherein two sets of physically coincident
nodes are placed along the interface, one set belonging to the
metal elements and the other to the mold elements. Heat transfer then occurs between these elements through a convective
relation, which is integrated along the interface. This is
opposed to the standard practice of a node belonging to both
metal and mold elements, which results in the assumption of
perfect conduction between them. Another possibility is to
insert a very thin element between mold and metal to represent
the die coat and air gap. Whether or not this would introduce
ill-conditioning to the system matrix by adding stiffness, and
how it compares numerically with the coincident node technique
are among the subjects of investigation.

Nomenclature

$\underline{C}$ = capacitance matrix

$f(t)$ = a function of time

h = convective heat transfer coefficient for boundary

H = enthalpy

k = conductivity

$\underline{K}$ = conductivity matrix

$\hat{n}$ = unit normal vector

N_i = shape function, weighting function

q = specified heat flux

t = time

$\hat{T}$ = temperature

T_a = ambient temperature

T_o = initial temperature

T = specified temperature

T_*^{n+1} = predicted temperature

$\underline{x}$ = vector of spatial coordinates

α = interface heat transfer coefficient

η = local coordinate

ξ = local coordinate

ρ = density

GOVERNING EQUATIONS

The application of the finite element method to the heat conduction equation has been treated in detail in many places [11,12,13], so only the outline will be sketched here. Conservation of energy yields the governing differential equation for the domain which, assuming temperature dependent thermal properties, becomes

$$\underline{\nabla}.(k \, \underline{\nabla} \, T) = \rho \, \frac{dH}{dT} \, \dot{T} \tag{1}$$

The boundary curve Γ of Ω is assumed to consist of two distinct segments Γ_1 and Γ_2.

Dirichlet and Neuman boundary conditions are specified on Γ_1 and Γ_2 respectively, i.e.

$$T(\underline{x},t) = \hat{T}(\underline{x})f(t) \text{ on } \Gamma_1 \tag{2}$$

$$k\underline{\nabla}T.\hat{n} + q + h(T-T_A) = 0 \text{ on } \Gamma_2 \tag{3}$$

For transient problems, the initial condition must also be specified in the form

$$T(\underline{x},0) = T_o(\underline{x}) \tag{4}$$

Discretizing the spatial domain with finite elements and employing the Galerkin form of the weighted residual method, the following system of ordinary differential equations is obtained:

$$\underline{C}\ \dot{\underline{T}} + \underline{K}\ \underline{T} = \underline{F} \tag{5}$$

where $\underline{T}$ denotes the vector of nodal values, T_i.

Typical components of the matrices $\underline{C}$ and $\underline{K}$ and the vector $\underline{F}$ are defined by

$$C_{ij} = \int_{\Omega} \rho \frac{dH}{dT} N_i N_j \ d\Omega \tag{6}$$

$$K_{ij} = \int_{\Omega} \underline{\nabla} N_j \cdot (k \ \underline{\nabla} N_i) \ d\Omega + \int_{\Gamma_2} h \ N_i N_j \ d\Gamma_2 \tag{7}$$

$$F_i = -\int_{\Gamma_2} (q - h \ T_A) \ N_i \ d\Gamma_2 \tag{8}$$

N_i is the shape function associated with node i, at which the temperature T_i is evaluated. Latent heat evolution is accounted for by the term dH/dT. In the current program, a smoothing function is used to determine the slope of the enthalpy curve at a particular temperature. Alternative procedures are discussed in [12].

A two step predictor-multicorrector scheme is utilized for marching in time, where the matrices are reformulated at each iteration. This scheme can be expressed as

Predictor:

$$[\underline{C}(\underline{T}^n) + \Delta t^n \ \theta \ \underline{K}(\underline{T}^n)] \underline{T}_*^{n+1} = [\underline{C}(\underline{T}^n) - \Delta t^n (1-\theta) \underline{K}(\underline{T}^n)] \underline{T}^n +$$

$$\Delta t^n \underline{F}(\underline{T}^n) \tag{9}$$

Corrector:

$$[\underline{C}(\underline{T}_p^n) + \Delta t^n \ \theta \ \underline{K}(\underline{T}_p^n)] \underline{T}_{p+1}^{n+1} = [\underline{C}(\underline{T}_p^n) - \Delta t^n (1-\theta) \underline{K}(\underline{T}_p^n)] \underline{T}^n +$$

$$\Delta t^n \underline{F}(\underline{T}_p^n) \tag{10}$$

where $\underline{T}_p^n = \beta \ \underline{T}_p^{n+1} + (1-\beta) \underline{T}^n \tag{11}$

and $\underline{T}_o^{n+1} = \underline{T}_*^{n+1} \tag{12}$

The superscript n refers to the time level and the subscript p refers to the number of corrective iterations.

A second order accurate, unconditionally stable Crank-Nicolson scheme is obtained with $\theta = \beta = 1/2$, [14]. The timestep is increased or decreased automatically depending upon the number of corrective iterations required to reach a specified convergence criterion.

COINCIDENT NODE TECHNIQUE

Consider two adjacent elements as in Figure 1. The nodal topology would be described as follows:

Element	Nodes			
1	1	3	4	2
2	3	5	6	4

When the system matrices are assembled, and because the elements have nodes 3 and 4 in common, a condition of perfect conduction between the elements will result.

If however, a second spatially coincident node is placed at each interface node, with one node belonging exclusively to one element as in Figure 2, a different topology will be obtained.

Element	Nodes			
1	1	3	4	2
2	5	7	8	6

If the matrices are assembled in the normal manner, a condition of perfect insulation will exist between the elements.

Convective-like heat transfer, i.e. where the flux is porportional to a temperature difference, between the elements may be effected by the following addition to the element conductivity matrix:

$$K^e = K^e + \sum_{i,j=1}^{M} \int_{\Gamma_3^e} \alpha \, N_i \, (N_j - 1/2 \, N_k) d\Gamma_3^e \qquad (13)$$

M is the number of nodes on the interface boundary Γ_3^e. and k is the number of the node coincident with j. This integration must be performed for the elements on both sides of the interface. Since both will contribute to the $N_i N_k$ crossterm in the global matrix, the factor 1/2 is required in (13).

Again with reference to Figure 2, the element topology must be expanded to account for the coincident nodes since terms involving them will now appear in the element matrix. Thus,

Element	Nodes					
1	1	3	4	2	5	6
2	5	7	8	6	3	4

Some sample integrations follow for the linear rectangular elements shown.

$$N_3 = N_5 = \frac{1}{2} (1-\xi) \tag{14}$$

$$N_4 = N_6 = \frac{1}{2} (1+\xi) \tag{15}$$

$$\gamma = \left| \left(\frac{\partial x}{\partial \xi}\right)^2 + \left(\frac{\partial y}{\partial \xi}\right)^2 \right|^{\frac{1}{2}}$$

$$= \left| \left(\sum_{i=1}^{M} \frac{\partial N_i}{\partial \xi} x_i\right)^2 + \left(\sum_{i=1}^{M} \frac{\partial N_i}{\partial \xi} y_i\right)^2 \right|^{\frac{1}{2}}$$

$$= \Delta y/2 \tag{16}$$

$$K^c_{33} = \int_{-1}^{1} \alpha\, N_3 N_3\, \gamma\, d\xi$$

$$= \frac{\alpha \Delta y}{8} \int_{-1}^{1} (1-\xi)^2\, d\xi$$

$$= \frac{\alpha \Delta y}{3} \tag{17}$$

$$K^c_{34} = \int_{-1}^{1} \alpha\, N_3 N_4\, \gamma\, d\xi$$

$$= \frac{\alpha \Delta y}{8} \int_{-1}^{1} (1-\xi^2)\, d\xi$$

$$= \frac{\alpha \Delta y}{6} \tag{18}$$

Summing the contributions to the crossterms from both elements yields

$$K^c_{35} = - K^c_{33} = - \frac{\alpha \Delta y}{3} \tag{19}$$

$$K^c_{36} = - K^c_{34} = - \frac{\alpha \Delta y}{6} \tag{20}$$

The superscript c indicates that these are the components of the terms due to the coincident node method.

In three dimensions, the integrations would be over the surface between the elements and, in the isoparametric formulation, would appear as

$$I_{ij} = \int_{-1}^{1} \int_{-1}^{1} \alpha \, N_i(\xi,\eta) \, [N_j(\xi,\eta) - \frac{1}{2} N_k(\xi,\eta)] \beta \, d\xi \, d\eta \tag{21}$$

where

$$\beta = \left| \left(\frac{\partial x}{\partial \xi} \frac{\partial y}{\partial \eta} - \frac{\partial y}{\partial \xi} \frac{\partial x}{\partial \eta} \right)^2 + \left(\frac{\partial x}{\partial \xi} \frac{\partial z}{\partial \eta} - \frac{\partial z}{\partial \xi} \frac{\partial x}{\partial \eta} \right)^2 + \right.$$

$$\left. \left(\frac{\partial y}{\partial \xi} \frac{\partial z}{\partial \eta} - \frac{\partial z}{\partial \xi} \frac{\partial y}{\partial \eta} \right)^2 \right|^{\frac{1}{2}} \tag{22}$$

THIN ELEMENTS

An alternative to the coincident node method is simply to use very thin elements between metal and mold, with a linear interpolation across the smaller dimension. In the context of a linear element mesh, no further considerations arise and, referring to Figure 3, sample terms of the thin element conductivity matrix are as follows:

$$N_3 = \frac{1}{4} (1-\eta) (1-\xi), \quad N_4 = \frac{1}{4} (1-\eta) (1+\xi), \quad N_5 = \frac{1}{4} (1+\eta) (1-\xi),$$

$$N_6 = \frac{1}{4} (1+\eta) (1+\xi) \tag{23}$$

$$\det J = \frac{\partial x}{\partial \eta} \frac{\partial y}{\partial \xi} - \frac{\partial y}{\partial \eta} \frac{\partial x}{\partial \xi} = \frac{\Delta x \Delta y}{4} \tag{24}$$

$$\frac{\partial N_i}{\partial x} = \frac{\partial N_i}{\partial \eta} \frac{\partial y}{\partial \xi} \frac{1}{\det J} - \frac{\partial N_i}{\partial \xi} \frac{\partial y}{\partial \eta} \frac{1}{\det J} = \frac{\partial N_i}{\partial \eta} \frac{\Delta y}{2} \frac{1}{\det J} \tag{25}$$

$$\frac{\partial N_i}{\partial y} = - \frac{\partial N_i}{\partial \eta} \frac{\partial x}{\partial \xi} \frac{1}{\det J} + \frac{\partial N_i}{\partial \xi} \frac{\partial x}{\partial \eta} \frac{1}{\det J} = \frac{\partial N_i}{\partial \xi} \frac{\Delta x}{2} \frac{1}{\det J} \qquad (26)$$

$$K^e_{33} = \int_{-1}^{1} \int_{-1}^{1} k \left| \left(\frac{\partial N_3}{\partial x}\right)^2 + \left(\frac{\partial N_3}{\partial y}\right)^2 \right| \det J \, d\xi \, d\eta$$

$$= \int_{-1}^{1} \int_{-1}^{1} k \left| \left(\frac{\partial N_3}{\partial \eta} \frac{\Delta y}{2}\right)^2 + \left(\frac{\partial N_3}{\partial \xi} \frac{\Delta x}{2}\right)^2 \right| \frac{1}{\det J} \, d\xi \, d\eta$$

$$= \frac{k}{16\Delta x \Delta y} \int_{-1}^{1} \int_{-1}^{1} \left| \Delta y^2 \, (1-\xi)^2 + \Delta x^2 \, (1-\xi)^2 \right| \, d\xi \, d\eta$$

$$= \frac{k}{3} \frac{\Delta y}{\Delta x} \left| 1 + \frac{\Delta x^2}{\Delta y^2} \right| \qquad (27)$$

Similarly,

$$K^e_{34} = \frac{k}{6} \frac{\Delta y}{\Delta x} \left| 1 - 2 \frac{\Delta x^2}{\Delta y^2} \right| \qquad (28)$$

$$K^e_{35} = - \frac{k}{6} \frac{\Delta y}{\Delta x} \left| 1 - \frac{\Delta x^2}{2\Delta y^2} \right| \qquad (29)$$

$$K^e_{36} = - \frac{k}{6} \frac{\Delta y}{\Delta x} \left| 1 + \frac{\Delta x^2}{\Delta y^2} \right| \qquad (30)$$

If it is desired to equate the convective heat transfer of the coincident node method to the conduction of thin elements, then

$$q = k \, \Delta T / \Delta x = \alpha \, \Delta T \qquad (31)$$

and $\quad \alpha = k / \Delta x \qquad (32)$

Substituting this value for α into equations 17-20 and comparing with equations 27-30, we find that

$$K^e_{33} = K^c_{33} \left(1 + \frac{\Delta x^2}{\Delta y^2} \right) \qquad (33)$$

etc. With typical dimensions of $\Delta y = 10$ and $\Delta x = 0.1$, it is seen that thin elements yield terms that are at the most 0.02% different from those of the coincident node method. It should be noted though that unless the thin elements are given a zero heat capacity, they will also contribute to the global capacitance matrix, unlike coincident nodes.

When quadratic elements are being used in the metal and mold, it is desirable to construct a linear-by-quadratic element to model the interface, as in Figure 4. Typical corner and mid-side node shape functions are as follows:

$$N_1 = \frac{1}{4} (1+\xi) (\eta^2-\eta) \qquad (34)$$

$$N_2 = \frac{1}{2} (1+\xi) (1-\eta^2) \qquad (35)$$

One might be led to suspect that the introduction of such thin elements into the mesh might significantly increase the stiffness of the system matrix $\underline{R} = \underline{C}^{-1}\underline{K}$, as opposed to using the coincident node method. Therefore, an investigation was made of the spectra for the eigenvalue problem $\underline{K}x = \underline{C} \lambda x$ for the two meshes shown in Figures 5 and 6, one involving coincident nodes and the other the linear-by-quadratic thin element described above. The dimensions and material properties were typical of the casting simulations described in the next section. The aspect ratio, $\Delta y/\Delta x$, of the thin element was 350. It was found that a maximum difference of 1.9% occurred between the values of the two spectra, which was for the largest eigenvalue. This indicates that there should be very little difference in the numerical behaviour of the two methods, and this was in fact verified by employing both techniques in complete casting simulation meshes.

NUMERICAL AND EXPERIMENTAL RESULTS

A thin rectangular plate, surmounted by a riser, was cast with an aluminium-silicon-copper alloy (BS.1490:LM4) in a gravity die made of Grade 17 grey iron. Chromel/alumel thermocouples were placed at regular intervals along the centerline of the cavity and in the die. With a pouring temperature of approximately 750°C, the temperature-time curves depicted in Figure 7 were obtained. This experimental work was conducted at British Non-Ferrous Metals Technology Center, Wantage, England.

This casting was modelled in two dimensions with a mesh of 45 quadratic isoparametric elements, 223 nodes, representing the cross-section of the die and table, as shown in Figure 8. Due to symmetry, only half of the cross-section was required. The width of the riser was reduced in the mesh so that the ratio of the riser modulus (volume/surface) to that of the plate was the same in 2-D as in 3-D. The material properties for the alloy and the iron die are given in Tables I and II, and were obtained from [17] and [18]. The heat capacity of the die adjacent to the plate was increased by 40%

to account for conduction through the plate in the third dimension. A condition of free convection was assumed on the external boundary, except at the line of symmetry which was perfectly insulated. Heat transfer coefficients were calculated from Nusselt number relationships given in [19], assuming laminar flow and constant wall temperatures:

Vertical die faces: $h = 2.3 \times 10^{-4}$ cal/sec-cm²-°C

Horizontal die faces: $h = 9.7 \times 10^{-5}$ cal/sec-cm²-°C

Horizontal melt face: $h = 4.9 \times 10^{-4}$ cal/sec-cm²-°C

Actually, this problem is fairly insensitive to the value of h. Doubling these coefficients produces less than a 2°C difference in the temperature of the die during the course of solidification.

Before pouring the casting, the cavity walls were sprayed with a proprietary die coat to an approximate thickness of 0.1 mm. This coat principally determines the interface heat transfer coefficient prior to air gap formation. For a vertical flat plate, an estimated value for the coefficient α was obtained from [18] of 7.5×10^{-2} cal/sec-cm²-°C. Further, it was advised that the air gap should form at 4.8 seconds, at which time α would be reduced by 71%. These were the parameters that were initially used with the coincident node method. Ultimately, it was found that the experimental temperature-time curves could be closely approximated by varying these parameters in different regions of the interface. The vertical wall of the plate was divided into four equal intervals, numbered from the bottom, and the final values of α for these interfaces, plus that of the riser, are given in Table III. Also shown are the times to air gap formation and the consequent reduction in α. The resulting numerical curves are depicted in Figure 9. Based upon the parameters required to attain these results, it would seem that the air gap does not form instantaneously over the plate, but rather progresses upwards from the bottom, where solidification begins. Also, it occurs much later in the riser. The value of α in the riser is roughly half that of the rest of the casting. This could be due to a thicker die coat there, which is normally the case.

This same casting was then modelled using thin quadratic-by-linear elements at the interface instead of coincident nodes. These elements had a width of 0.1 mm with a typical aspect ratio of 120. The results differed by less than 0.05% from those of the coincident node method. However, because of the increased number of elements, the matrix assembly time was increased by 34% per iteration, which in this case represents a 21% increase in the total cpu time per iteration. When a problem is non-linear and the matrices have to be re-calculated at each iteration, this adds significantly to the cost of the solution. Thus, the coincident node method is to be preferred from the viewpoint of economy. It offers, as well, some advantages in mesh design since no nodal coordinates need be adjusted to form an interface. Thin elements, on the other hand, do not require any reprogramming, except perhaps to provide a linear-by-quadratic element.

Finally, a plate with a centered circular boss on one side was cast, with thermocouple measurements yielding the temperature-time curves shown in Figure 10. It is seen that the boss region solidifies after the top of the plate, indicating likely problems of shrinkage porosity. Because the casting is now asymmetric, the full cross-section must be enmeshed for the model (Figure 11). Using the heat transfer coefficient data in Table III with the coincident node method in this simulation gave the results shown in Figure 12, wherein the delayed solidification of the boss is evident. The 575°C contour at 12 seconds is illustrated in Figure 13. With the assumed latent heat evolution, this represents a fraction solidified of 80%, at which point feeding has probably ceased. Quite clearly, this zone has bridged across above the boss, thereby predicting a region of underfeeding with consequent porosity.

CONCLUSIONS

Two alternative techniques have been presented to deal with the interface between metal and mold when the finite element method is applied to casting simulation. Numerically, they are virtually equivalent. The coincident node method offers significant savings in computer time and greater ease of mesh design. This would be the method of choice if casting simulation was to be done on a regular basis, because then the reprogramming effort would be justified. Otherwise, thin elements could be used. Finally, it should be noted that these methods can be utilized in any situation where a thermal break occurs in a continuum, such as at parting faces in a die.

TABLE I   Material properties for BS.1490:LM4

Composition:

Element	Weight Percent
Aluminium	89.41
Silicon	5.70
Copper	3.00
Iron	0.70
Manganese	0.37
Zinc	0.32
Titanium	0.20
Nickel	0.16
Magnesium	0.06
Lead	0.05
Tin	0.03

Solidification Range: 625–525°C

Latent Heat: 92.3 cal/g

Density: 2.67 g/cm^3

Conductivity and Specific Heat as Functions of Temperature:

Temperature (°C)	Conductivity (cal/sec-cm-°C)	Specific Heat (cal/g-°C)
225	0.393	0.234
325	0.400	0.247
425	0.390	0.268
525	0.390	0.301
625	0.442	0.301
725	0.442	0.301

TABLE II   Material properties for Grade 17 Grey Iron

Composition:

Element	Weight Percent
Iron	93.55
Carbon	3.18
Silicon	2.03
Manganese	0.56
Phosphorus	0.59
Sulphur	0.09

Density: 7.23 g/cm³

Conductivity and Specific Heat as Functions of Temperature:

Temperature (°C)	Conductivity (cal/sec-cm-°C)	Specific Heat (cal/g-°C)
225	0.118	0.13
325	0.101	0.14
425	0.092	0.15
525	0.089	0.16
625	0.084	0.18
675	0.082	0.19

TABLE III   Interface Heat Transfer Coefficient Data

Interval	α(cal/sec-cm^2-°C)	Air Gap Formation Time (sec)	Reduction in α (%)
1	5.6×10^{-2}	4.3	80
2	4.9×10^{-2}	5.9	87
3	4.2×10^{-2}	6.1	87
4	4.9×10^{-2}	6.5	58
Riser	2.8×10^{-2}	27.5	70

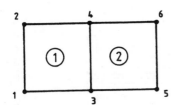

Figure 1   Linear elements
            with common
            interface nodes

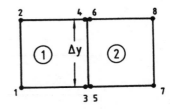

Figure 2   Linear elements
            with coincident
            interface nodes

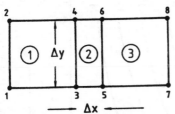

Figure 3   Linear elements
            with thin element
            at interface

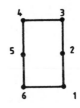

Figure 4   Linear-by-
            quadratic element

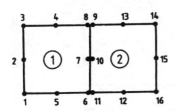

Figure 5   Quadratic elements
            with coincident
            node interface

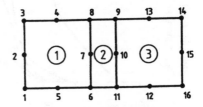

Figure 6   Quadratic
            elements with
            thin element
            interface

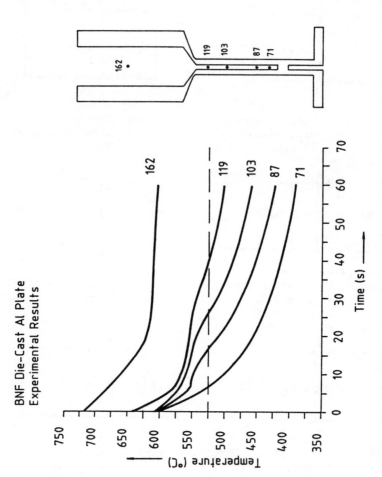

Figure 7 Experimental temperature-time curves of plate without boss

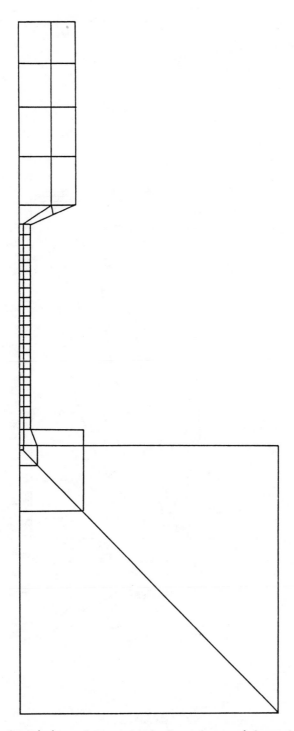

Figure 8   Finite element mesh for plate without boss

348

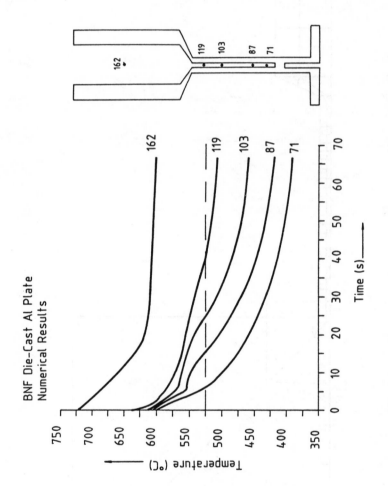

Figure 9   Numerical temperature-time curves of plate without boss

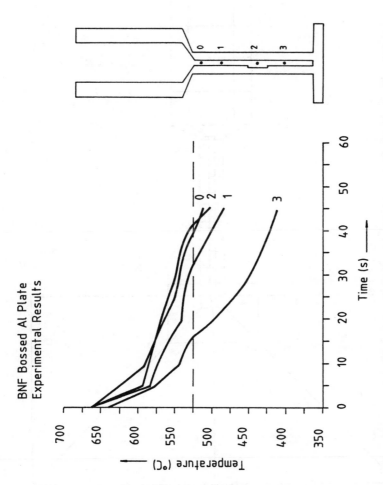

Figure 10 Experimental temperature-time curves of plate with boss

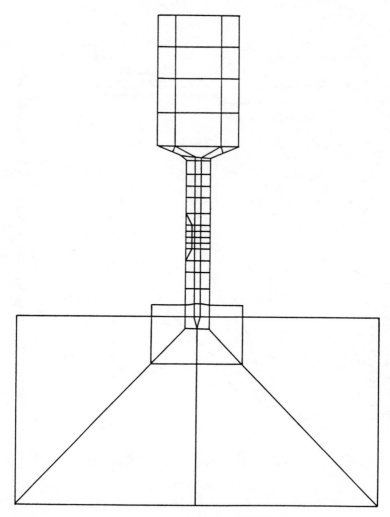

Figure 11 Finite element mesh for plate with boss

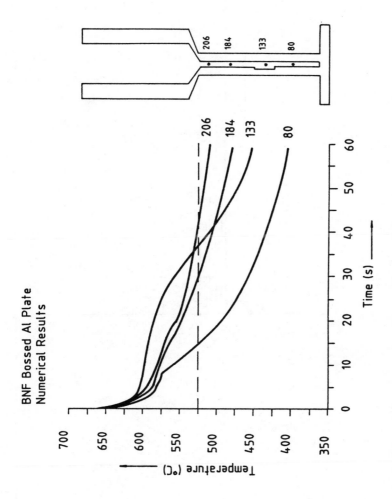

Figure 12 Numerical temperature-time curves of plate with boss

352

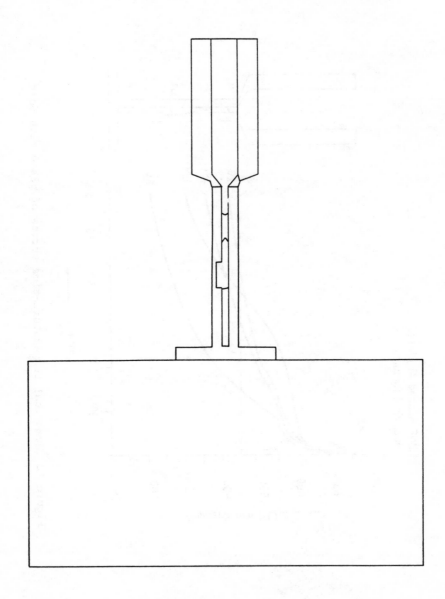

Figure 13 Eighty percent fraction solidified contour
at 12 seconds for plate with boss

REFERENCES

1. W.C. ERICKSON, "Computer Simulation of Solidification", AFS Int. Cast. Metals J., March 1980, pp. 30-40.

2. J.G. HENZEL, J. KEVARIAN, "Predicting Casting Solidification Patterns with a Computer", Foundry, May 1964, pp. 50-53.

3. A. JEYARAJAN, R.D. PEHLKE, "Casting Design by Computer", AFS Trans., Vol. 101, 1975, pp. 405-412.

4. P.R. SAHM, "Applications-Properties-Microstructure-Solidification Technologies: Challenge for Computer Simulation and Modelling Foundry Processes", Solidification Processes: Computer Simulation and Modelling Workshop, CIATF - Cairo, Nov. 1983, pp. 2-18.

5. N.G. SEMAN, "Foundries are Closing the Computer Software Gap - Parts I and II", Foundry M&T, Sept. and Oct. 1981.

6. R. SUN, "Simulation and Study of Surface Conductance for Heat Flow in the Early Stages of Casting", AFS Cast Metals Res. J., Vol. 6, No.3, 1970, pp. 105-110.

7. M. PRATES, H. BILONI, "Variables Affecting the Nature of the Chill Zones", Metallurgical Trans., Vol.3, 1972, pp. 1501-1510.

8. J.W. GRANT, "Thermal Modelling of a Permanent Mold Casting Cycle", Modelling of Casting and Welding Processes, AIME, Warrendale, PA, 1981, pp. 19-37.

9. O.K. RIEGGER, "Application of a Solidification Model to the Die Casting Process", ibid, pp. 39-72.

10. A. JEYARAJAN, R.D. PEHLKE, "Computer Simulation of Solidification of a Casting with a Chill", AFS Trans., Vol. 84, 1976, pp. 647-52.

11. O.C. ZIENKIEWICZ, K. MORGAN, "Finite Elements and Approximation", John Wiley, New York, 1983.

12. K. MORGAN, R.W. LEWIS, O.C. ZIENKIEWICZ, "An Improved Algorithm for Heat Conduction Problems with Phase Change", Int. J. Num. Meth. Engng., Vol.12, 1978, pp. 1191-1195.

13. O.C. ZIENKIEWICZ, The Finite Element Method, 3rd Edition, McGraw-Hill, 1977.

14. T.J.R. HUGHES, "Unconditionally Stable Algorithms for Nonlinear Heat Conduction", Comp. Meth. in App. Mech. and Engng., Vol.10, 1977, pp. 135-139.

15. L.J.D. SULLY, "The Thermal Interface Between Castings and Chill Molds", Trans. AFS, Vol. 84, 1976, pp. 735-744.

16. C.W. NELSON, "Wärmeübergang an der Formwand beim Druckgiessen", Giesserei Praxis, No.19, 1972, pp. 341-349.

17. Thermophysical Properties of High Temperature Solid Materials, Vol.3, Thermophysical Properties Research Center, Purdue University, Collier MacMillan, London, 1967.

18. Personal communication from British Non-Ferrous Metals Technology Center, Wantage, England.

19. M.N. OZISIK, Basic Heat Transfers, McGraw-Hill, New York, 1977, pp. 303-309.

SUBJECT INDEX